Three Mile Island

Three Mile Island explains the far-reaching consequences of the partial meltdown of Pennsylvania's Three Mile Island Power on March 28, 1979. Though the disaster was ultimately contained, the fears it triggered had an immediate and lasting impact on public attitudes towards nuclear energy in the United States. In this volume, Grace Halden contextualizes the events at Three Mile Island and the ensuing media coverage, offering a gripping portrait of a nation coming to terms with technological advances that inspired both awe and terror. Including a selection of key primary documents, this book offers a fascinating resource for students of the history of science, technology, the environment, and Cold War culture.

Grace Halden is Lecturer in Modern and Contemporary Literature at Birkbeck College, University of London.

Critical Moments in American History

Edited by William Thomas Allison, Georgia Southern University

A full list of titles in this series is available at: https://www.routledge.com/Critical-Moments-in-American-History/book-series/CRITMO.

Recently published titles:

The Battle of Fort Sumter
The First Shots of the American Civil War
Wesley Moody

The WPA
Creating Jobs and Hope in the Great Depression
Sandra Opdycke

The California Gold Rush
The Stampede that Changed the World
Mark Eifler

Bleeding Kansas
Slavery, Sectionalism, and Civil War on the Missouri-Kansas Border
Michael E. Woods

The Marshall Plan
A New Deal for Europe
Michael Holm

The Espionage and Sedition Acts
World War I and the Image of Civil Liberties
Mitchell C. Newton-Matza

McCarthyism
The Realities, Delusions and Politics Behind the 1950s Red Scare
Jonathan Michaels

Three Mile Island

The Meltdown Crisis and Nuclear Power in American Popular Culture

Grace Halden

NEW YORK AND LONDON

First published 2017
by Routledge
711 Third Avenue, New York, NY 10017

and by Routledge
2 Park Square, Milton Park, Abingdon, Oxon OX14 4RN

Routledge is an imprint of the Taylor & Francis Group, an informa business

Library of Congress Cataloging in Publication Data
Names: Halden, Grace, 1983– author.
Title: Three Mile Island: the meltdown crisis and nuclear power in American popular culture/by Grace Halden.
Description: New York: Routledge, 2017. | Series: Critical moments in American history | Includes bibliographical references and index.
Identifiers: LCCN 2016049697 (print) | LCCN 2016051035 (ebook) | ISBN 9781138917637 (hardback) | ISBN 9781138917644 (paperback) | ISBN 9781315688039 (electronic) | ISBN 9781315688039
Subjects: LCSH: Three Mile Island Nuclear Power Plant (Pa.) | Nuclear power plants—Pennsylvania—Accidents. | Nuclear power plants—Accidents—Social aspects—United States. | Nuclear engineering—United States—Public opinion.
Classification: LCC TK1345.H37 H35 2017 (print) | LCC TK1345.H37 (ebook) | DDC 363.17/99—dc23
LC record available at https://lccn.loc.gov/2016049697

ISBN: 978-1-138-91763-7 (hbk)
ISBN: 978-1-138-91764-4 (pbk)
ISBN: 978-1-315-68803-9 (ebk)

Typeset in Bembo & Helvetica Neue
by Florence Production Ltd, Stoodleigh, Devon, UK

For David, Joy, and James Halden.

You have my love and thanks.

Contents

Series Introduction

Welcome to the Routledge *Critical Moments in American History* series. The purpose of this new series is to give students a window into the historian's craft through concise, readable books by leading scholars, who bring together the best scholarship and engaging primary sources to explore a critical moment in the American past. In discovering the principal points of the story in these books, gaining a sense of historiography, following a fresh trail of primary documents, and exploring suggested readings, students can then set out on their own journey, to debate the ideas presented, interpret primary sources, and reach their own conclusions—just like the historian.

A critical moment in history can be a range of things—a pivotal year, the pinnacle of a movement or trend, or an important event such as the passage of a piece of legislation, an election, a court decision, a battle. It can be social, cultural, political, or economic. It can be heroic or tragic. Whatever they are, such moments are by definition "game changers," momentous changes in the pattern of the American fabric, paradigm shifts in the American experience. Many of the critical moments explored in this series are familiar; some less so.

There is no ultimate list of critical moments in American history—any group of students, historians, or other scholars may come up with a different catalog of topics. These differences of view, however, are what make history itself and the study of history so important and so fascinating. Therein can be found the utility of historical inquiry—to explore, to challenge, to understand, and to realize the legacy of the past through its influence of the present. It is the hope of this series to help students realize this intrinsic value of our past and of studying our past.

William Thomas Allison
Georgia Southern University

Figures

Acknowledgments

The people who kindly and selflessly contributed to this project are numerous.

Thank you to those who have helped me with research, including The British Library, Adam Scott Clark, Dan Kelton, Library of Congress, Phil Noble (EDF), Brendan Olivera (and Sasquesahanough Lodge 11), James Ollinger, Elizabeth Rosa, Amanda Rose, Robert Del Tredici, and Colin Tucker (EDF). Recognition also for Theodore Meyer for helping me prepare the book for publication.

The support of Birkbeck College, University of London, has been exceptional. With special mention to Anthony Bale, Heike Bauer, Beatrice Bazell, Mark Blacklock, Nicola Bown, Joseph Brooker, Isabel Davis, Harriet Earle, Martin Eve, Anna Hartnell, Roger Luckhurst, Louise Owen, Ana Parejo Vadillo, Anthony Shepherd, Tony Venezia, Sarah Walker, Carol Watts, Sue Wiseman, and Gillian Woods.

With much love to my family and friends who have always supported me—even when it means seeing so little of me and reading endless drafts. With special thanks to Lee Smith, whose contribution and assistance has been absolutely invaluable. With thanks to Joy, David, and James for endless support and guidance.

Timeline: Three Mile Island in Context

This timeline is not exhaustive, but it provides an overview of some major and influential moments in the twentieth and twenty-first centuries that relate to the event at Three Mile Island.

1933–1945	**Significant Events**: President Franklin D. Roosevelt is President of the United States.
1942	**Nuclear Context**: The Manhattan Project is formed to create the world's first nuclear bomb. The first successful, controlled nuclear chain reaction is achieved by Enrico Fermi's team at the University of Chicago (Chicago Pile-1).
1945–1953	**Significant Events**: President Harry S. Truman is President of the United States.
1945	**Nuclear Context**: Following successful testing in Alamogordo, New Mexico (Trinity Test, July 16), Japan becomes the first victim of the atomic bomb when Little Boy detonates on Hiroshima on August 6, and Fat Man detonates on Nagasaki on August 9.
1945	**Nuclear Context**: Manhattan Project scientists start the *Bulletin of the Atomic Scientists*. The newsletter/magazine is created because the scientists involved in atomic development "could not remain aloof to the consequences of their work".[1]
1946	**Nuclear Context**: The United States begins testing nuclear weapons at Bikini Atoll (1946–1958). Nuclear testing dominates the following decades.
1946	**Nuclear Context**: President Harry S. Truman signs the Atomic Energy Act (August 1, 1946), which allows atomic energy to fall under civilian control. The United States Atomic Energy Commission (AEC) is formed to help secure peaceful atomic power. The Joint Committee on Atomic Energy (JCAE) is formed.
1947	**Cultural Context**: *The Beginning or the End* is accredited as the first film to document the development of the nuclear bomb.
1947	**Nuclear Context**: The *Bulletin of the Atomic Scientists* adds 'The Doomsday Clock' to the front cover in 1947. To reach midnight is to reach disaster. The time is seven minutes to midnight.
1949	**Significant Events**: North Atlantic Treaty Organization (NATO) is established (April 4).
1949	**Nuclear Context**: The Soviet Union's first atomic bomb test, 'First Lightning' (known as 'Joe-1' in America).

1949	**Nuclear Context**: The Doomsday Clock is set at three minutes to midnight following the Soviet Union's first nuclear test.
1950	**Cultural Context**: *The Motorola Television Hour* shows the film 'Atomic Attack', based on the Judith Merril story *Shadow on the Hearth* (1950). The story features a family at home during a nuclear crisis and a child becoming seriously ill from radiation contamination.
1951	**Nuclear Context**: The United States begins nuclear testing in Nevada. The Nevada Test Site would become known as the "most bombed place on Earth".[2]
1951	**Nuclear Context**: Experimental Breeder Reactor 1 (EBR-1) at Idaho National Engineering and Environmental Laboratory becomes the United States' first electricity producing nuclear reactor. The reactor generates usable electricity and lights four light bulbs.
1952	**Cultural Context**: The film *Atomic City* deals with terrorists threatening a nuclear physicist.
1952	**Nuclear Context**: The United Kingdom tests its first atomic bomb.
1952	**Nuclear Context**: At Enewetak (Eniwetok) Atoll, the first hydrogen bomb is tested by the United States.
1952	**Nuclear Accidents**: A severe nuclear accident occurs at the NRX reactor near Chalk River, Canada (commonly known as the Chalk River nuclear accident).
1953–1961	**Significant Events**: Dwight D. Eisenhower is President of the United States.
1953	**Significant Events**: Julius and Ethel Rosenberg are executed for sharing secret atomic information with the Soviet Union.
1953	**Nuclear Context**: The Doomsday Clock is set at two minutes to midnight to reflect thermonuclear development.
1953	**Nuclear Context**: In December, President Dwight D. Eisenhower delivers his famous 'Atoms for Peace' speech at the United Nations' General Assembly.
1954	**Nuclear Context**: At Bikini Atoll, islanders and local islands, as well as a Japanese fishing boat, are contaminated by fallout after tests.
1954	**Nuclear Context**: Atomic Energy Act of 1954 demonstrates amendments from the earlier version. It declares that:

> Atomic energy is capable of application for peaceful as well as military purposes. It is therefore declared to be the policy of the United States that—
>
> a. the development, use, and control of atomic energy shall be directed so as to make the maximum contribution to the general welfare, subject at all times to the paramount objective of making the maximum contribution to the common defense and security; and
>
> b. the development, use, and control of atomic energy shall be directed so as to promote world peace, improve the general welfare, increase the standard of living, and strengthen free competition in private enterprise.[3]

1954	**Nuclear Context**: The Soviet Union's Obninsk Nuclear Power Plant generates electricity and is connected to the external power grid.
1955	**Nuclear Context**: The small settlement of Arco, Idaho becomes the first American town to be powered by nuclear energy (from Borax III).

1955 **Nuclear Context**: In Switzerland, the International Conference on the Peaceful Uses of Atomic Energy begins.

1955 **Nuclear Context**: The Soviet Union successfully tests its first true fusion weapon.

1955 **Nuclear Context**: EBR-1 experiences partial meltdown associated with human error.

1956 **Nuclear Context**: The first full-scale nuclear power plant is opened at Calder Hall. Queen Elizabeth II, who opened the plant, states, "This new power, which has proved itself to be such a terrifying weapon of destruction, is harnessed for the first time for the common good of our community."[4]

1957 **Nuclear Context**: The AEC's WASH-740, 'Theoretical Possibilities and Consequences of Major Accidents in Large Nuclear Power Plants' (otherwise known as The Brookhaven Report) is published. The report looks at the potential for nuclear power accidents and concludes that the worst feasible accident could result in 3,400 deaths.[5] Many consider this report to have been commissioned and diluted to "reassure the public" that anxieties are baseless.[6] Later reports quote much higher level of fatalities but these reports are suppressed.[7] See also the Rasmussen Report (WASH-1400), 1975.

1957 **Nuclear Context**: The United Kingdom starts nuclear testing in the Pacific Ocean at Christmas Island.

1957 **Nuclear Context**: The International Atomic Energy Agency (IAEA) is established.

1957 **Nuclear Context**: The origins of AEC's Project Plowshare.

1957 **Nuclear Accidents**: Although details are suppressed, a serious nuclear incident occurs in the Ural Mountains in the Soviet Union.

1957 **Nuclear Context**: The Price-Anderson Act is introduced. The Act aims to "cover liability claims of members of the public for personal injury and property damage caused by a nuclear accident involving a commercial nuclear power plant".[8] The hope is that by limiting utility financial liability following a nuclear accident, more plants will be constructed.

1957 **Nuclear Accidents**: A fire breaks out at the Rocky Flats plant, Colorado. Another incident occurs in 1969.

1957 **Nuclear Accidents**: A fire at Windscale No 1 in Cumberland (now Cumbria), England, releases a radioactive cloud; this incident becomes the largest and most severe nuclear disaster in the history of the United Kingdom.

1957 **Nuclear Context**: Shippingport, Pennsylvania becomes the first large-scale nuclear power plant to produce commercial power purely for peaceful purposes.

1958 **Nuclear Context**: Euratom (European Atomic Energy Community) is formed to assist atomic development in Europe.

1960 **Nuclear Context**: The Doomsday Clock is set at seven minutes to midnight.

1961–1963 **Significant Events**: John F. Kennedy is the President of the United States.

1961 **Nuclear Accidents**: The United States experiences a major nuclear reactor accident at SL-1 (Stationary Low Power Plant 1) when three servicemen are killed at Idaho Falls.

1961 **Significant Events**: Major Yuri Gagarin becomes the first man in space, reflecting a major step forward in the Space Race by the Soviet Union.

1962 **Nuclear Context**: The Cuban Missile Crisis lasts thirteen days in October and involves a tense confrontation between Cold War powers the United States and the Soviet Union over nuclear weapon placement.

1963–1969	**Significant Events**: Lyndon B. Johnson is President of the United States.
1963	**Nuclear Context**: In Moscow, a Limited Test Ban Treaty is signed barring atmospheric nuclear testing, as well as testing under water and in space.
1963	**Nuclear Context**: The Doomsday Clock is set back to twelve minutes to midnight. This reflects calming tensions following the Partial Test Ban Treaty.
1964	**Cultural Context**: Stanley Kubrick's film *Dr Strangelove or: How I Learned to Stop Worrying and Love the Bomb* deals satirically with Cold War tensions and nuclear conflict.
1965	**Nuclear Context**: Following the Soviet Union's lead in launching nuclear reactors into space, the United States launches SNAP-10A.
1966	**Nuclear Accidents**: The United States Air Force accidentally drops four nuclear bombs over Spain.
1966	**Nuclear Accidents**: The fuel core of the Enrico Fermi experimental breeder reactor partially melts.
1967	**Three Mile Island**: In Pennsylvania, on the Susquehanna River's Three Mile Island, construction starts on the first of two core reactors, 'Unit-1'.
1968	**Nuclear Context**: Treaty on the Non-Proliferation of Nuclear Weapons adopted (ratified 1970).
1968	**Nuclear Context**: The Doomsday Clock is moved forward again to seven minutes to midnight due to nuclear advancements in other countries and a range of conflicts.
1969–1974	**Significant Events**: Richard Nixon is President of the United States.
1969	**Cultural Context**: America puts the first man on the moon as Neil Armstrong takes a "giant leap for mankind".
1969	**Nuclear Context**: The Doomsday Clock is moved back three minutes and reads ten minutes to midnight following the Nuclear Non-Proliferation Treaty. Part of the agreement is to advance nuclear power over nuclear armaments.
1970	**Three Mile Island**: Three Mile Island sees the construction of the second core reactor, 'Unit-2'.
1970	**Cultural Context**: Earth Day on April 22 sees twenty million Americans protest against damage to the environment and speak out about their concerns over the treatment of the planet.
1970	**Significant Events**: Congress creates the Environmental Protection Agency following increasing concern over pollution and environmental issues.
1972	**Nuclear Context**: Strategic Arms Limitation Treaty (SALT) and the Anti-Ballistic Missile Treaty (ABM) are signed by the Soviet Union and the United States. Further limitations are discussed as part of negotiations for SALT II; SALT II is signed in 1979.
1972	**Nuclear Context**: The Doomsday Clock is moved back again to twelve minutes to midnight following treaty successes.
1973	**Significant Events**: Although energy problems occur before this date, in 1973 the 'energy crisis' becomes an international issue and recession worsens due to many factors. One contributing problem is a dramatic increase in oil prices.
1974–1977	**Significant Events**: Following the resignation of Richard Nixon, Gerald Ford takes over as President of the United States.

1974 **Nuclear Context**: Threshold Test Ban Treaty (TTBT) is signed, limiting test yields.

1974 **Nuclear Context**: President Gerald Ford establishes The Energy Reorganization Act, which replaces the AEC with the Nuclear Regulatory Commission (NRC) and the Energy Research and Development Administration (ERDA).

1974 **Nuclear Context**: Karen Silkwood dies in a car crash on her way to discuss problems at Kerr-McGee with *The New York Times*.

1974 **Three Mile Island**: Three Mile Island Unit-1 (TMI-1), a pressurized-water reactor, starts commercial operation.

1974 **Nuclear Context**: The Doomsday Clock is moved forward and reads nine minutes to midnight as nuclear weapons development continues.

1975 **Nuclear Context**: Twenty-three nuclear reactors are shut down because of cracked coolant pipes by the newly established NRC.

1975 **Nuclear Context**: The Rasmussen Report (WASH-1400), directed by Norman Rasmussen for the NRC (formerly initiated by the AEC), seeks to analyze health and economic risks from potential severe accidents at PWRs (Pressurized Water Reactor) and BWRs (Boiling Water Reactors). The report is met with mixed responses; and, although very influential, many are critical as to the report's reliability.

1975 **Nuclear Accidents**: The Tennessee Valley Authority's Browns Ferry Nuclear Power Plant in Alabama experiences a crisis when a worker searching for air leaks with a candle inadvertently starts a fire which spreads and lasts for hours.

1975 **Cultural Context**: Based on the 1966 Enrico Fermi accident, *We Almost Lost Detroit* by John G. Fuller is published.

1976 **Nuclear Accidents**: Workers are contaminated after an explosion at Hanford Nuclear Reservation. Due to high levels of radiation contamination, one worker, Harold McCluskey, is dubbed 'Atomic Man'.

1976 **Nuclear Context**: Whistleblowers Gregory C. Minor, Richard B. Hubbard, and Dale G. Bridenbaugh speak out against safety problems at General Electric plants. All three resign as engineers for GE. Minor claims his resignation is due to feeling that nuclear power poses a significant risk to human health.[9]

1977–1981 **Significant Events**: Jimmy Carter is President of the United States. Carter, who has a background in nuclear physics, is President during the Three Mile Island crisis.

1978 **Three Mile Island**: At the dedication ceremony for Three Mile Island in 1978, John F. O'Leary of the Energy Department reportedly declares the plant to be "a bright and shining option for this country . . . the best option available to us".[10] Three Mile Island Unit-2 (TMI-2) starts operation.

1979 **Cultural Context**: James Bridges' *The China Syndrome* is released.

1979 **Three Mile Island**: Disaster at Three Mile Island

March 28 After only a few months of operation, Unit-2 experiences a partial core meltdown as a result of human error and equipment failure. During the first day of the crisis, conflicting reports over the severity of the crisis are issued. Governor William Scranton Jr concedes that the problems at the plant are more complicated than initial reports released by Metropolitan Edison suggest.

March 30 Governor Dick Thornburgh calls for residents to remain in their homes with windows and doors securely fastened. Pregnant women and young children in a five-mile radius are

	advised to evacuate but much higher numbers flee the area. Concerns are expressed over earlier reports of a hydrogen bubble.
March 31	Continuing fears in the press of a hydrogen bubble forming in the pressure-vessel make many worry about containment failure.
April 1	President Jimmy Carter visits the plant.
April 4	Governor Thornburgh announces that the crisis is over on the *Today* show.
April 9	The crisis is declared to be at an end by Harold Denton from the NRC. Thornburgh announces that evacuees can return and schools will reopen on April 10.
1979	**Nuclear Context**: The Silkwood estate is awarded $10 million in punitive damages as a result of inadequate health and safety at Kerr-McGee that caused Karen Silkwood's contamination.
1979	**Nuclear Context**: Carter announces his intent to increase funds for solar energy development.
1979	**Nuclear Context**: SALT II Treaty is signed by Carter and Leonid Brezhnev, securing, among many things, a limitation on the further development of nuclear weapons. However, the United States Senate chooses not to ratify the agreement.
1980	**Nuclear Context**: The Energy Security Act is signed by Carter.
1980	**Three Mile Island**: The plant prepares for cleanup. Radioactive gas is vented. The first manned entry into the reactor. Talks are held over whether Unit-1 should ever be restarted. A meeting is held by the Advisory Panel for the Decontamination of Unit-2—the meeting sees a mixture of experts and citizens in attendance.
1980	**Nuclear Context**: The Doomsday Clock reads seven minutes to midnight.
1981–1989	**Significant Events**: Ronald Reagan is President of the United States.
1981	**Nuclear Context**: The Doomsday Clock reads four minutes to midnight; a leap forward that highlights growing Cold War tensions.
1983	**Nuclear Context**: The issue of nuclear waste comes to the forefront and the Nuclear Waste Policy Act of 1982 is enacted.
1983	**Cultural Context**: The film *Silkwood* is released in movie theatres.
1984	**Nuclear Context**: The Doomsday Clock reads three minutes to midnight.
1985	**Three Mile Island**: After protests and much debate spanning from 1979 to 1985, Unit-1 is restarted.
1986	**Nuclear Accidents**: A catastrophic nuclear accident occurs at the Chernobyl plant in the Ukraine.
1986	**Nuclear Context**: An out of court settlement is reached in the Silkwood case for $1.38 million.
1987	**Nuclear Context**: Intermediate-Range Nuclear Forces Treaty is agreed, which seeks to reduce nuclear arsenals and eliminate short- and medium-range nuclear weapons.
1988	**Nuclear Context**: The Doomsday Clock reads six minutes to midnight, showing an easing of Cold War tensions as the Intermediate-Range Nuclear Forces Treaty is agreed.
1989–1993	**Significant Events**: George H. W. Bush is President of the United States.
1990	**Three Mile Island**: Fuel removal is completed.
1990	**Nuclear Context**: Although nearly completed, a nuclear power plant in Michigan is converted to gas due to nuclear opposition and economic concerns. Economic concerns

would continue to plague the industry, and by 1996 six plants would close because of economic factors.[11]

1990 **Nuclear Context**: The Doomsday Clock reads ten minutes to midnight as the Cold War reaches its end.

1991 **Three Mile Island**: Water generated during the accident is processed. By 1993, 2.23 million gallons of water are processed.

1991 **Nuclear Context**: The Strategic Arms Reduction Treaty (START) is signed by Bush and Mikhail Gorbachev, which seeks to reduce nuclear warheads.

1991 **Nuclear Context**: For the first time, the Doomsday Clock reads seventeen minutes to midnight. The Cold War is over.

1993–2001 **Significant Events**: Bill Clinton is President of the United States.

1993 **Three Mile Island**: An intruder hides inside the Three Mile Island complex after invading the site by crashing through gates in a car.

1993 **Three Mile Island**: The last meeting is held for The Advisory Panel for Decontamination of TMI-2. Monitored storage starts. Cleanup concludes.

1995 **Nuclear Context**: Based on American pressurized water reactor technology, the United Kingdom's Sizewell B PWR starts generation.

1995 **Nuclear Context**: The Doomsday Clock moves from seventeen to fourteen minutes to midnight. The end of the Cold War does not signal nuclear disbandment.

1996 **Nuclear Accident**: A valve leak at Waterford, Connecticut causes shutdowns.

1997 **Nuclear Accidents**: An accident at Tokaimura nuclear fuel-processing plant in Japan causes the contamination of thirty-five workers. Another accident contaminates workers in 1999.

1998 **Nuclear Context**: The Doomsday Clock reads nine minutes to midnight as other nations test nuclear weapons. Nuclear development continues.

2000 **Nuclear Context**: Chernobyl is completely shut down.

2001–2009 **Significant Events**: George W. Bush is President of the United States.

2001 **Significant Events**: On September 11, a series of coordinated attacks against America are conducted by the terrorist group al-Qaeda.

2002 **Nuclear Accidents**: Cavities are discovered in the rector pressure vessel at Davis-Besse, Ohio. They were undetected for years. The plant is shut down for two years.

2002 **Nuclear Context**: The Doomsday Clock is set to seven minutes to midnight as concerns over terrorism increase.

2004 **Nuclear Accidents**: In Japan, a pipe bursts at Mihama Nuclear Power Plant and kills several maintenance workers.

2005 **Nuclear Context**: International Convention on the Suppression of Acts of Nuclear Terrorism is adopted.

2005 **Nuclear Context**: President George W. Bush signs the Energy Policy Act which seeks to support nuclear power development and homogenize the licensing and operation of nuclear plants.

2007 **Nuclear Context**: America's nuclear power industry experiences resurgence as new applications are made for new reactors following decades of slow development after 1979.

2007	**Nuclear Context**: The Doomsday Clock reads five minutes to midnight as nuclear testing continues, this time in North Korea.
2009–2017	**Nuclear Context**: Barack Obama is President of the United States. During Obama's presidency, he receives the Nobel Peace Prize partly for his commitment to a world free of nuclear weapons.
2010	**Nuclear Context**: The Doomsday Clock reads six minutes to midnight.
2011	**Nuclear Accidents**: After an earthquake and tsunami, the Fukushima Nuclear Power Plant in Japan experiences meltdowns during a catastrophic crisis.
2012–2016	**Nuclear Context**: The Doomsday Clock moves from five to three minutes to midnight. This reflects a range of pressing issues from concerns over dangerous reactors to climate change.[12]

NOTES

1 *Bulletin of the Atomic Scientists*, 'Background and Mission: 1945–2016', *thebulletin.org* (2016), available online at http://thebulletin.org/background-and-mission-1945-2016 (accessed August 8, 2016).
2 Topham, Laurence, Alok Jha, and Will Franklin, 'Building the bomb', *The Guardian* (September 22, 2015), available online at www.theguardian.com/us-news/ng-interactive/2015/sep/21/building-the-atom-bomb-the-full-story-of-the-nevada-test-site (accessed August 8, 2016).
3 'The Atomic Energy Act', United States Nuclear Regulatory Commission, 30 August 1954, Chapter 1, Sec. 1.
4 '1956: Queen Switches On Nuclear Power', *BBC News* (October 17, 1956), available online at http://news.bbc.co.uk/onthisday/hi/dates/stories/october/17/newsid_3147000/3147145.stm (accessed August 14, 2016).
5 Anna Gyorgy, *No Nukes: Everyone's Guide to Nuclear Power* (Montreal: Black Rose Books, 1979), p. 111.
6 Harry Henderson, *Nuclear Power: A Reference Handbook*, 2nd edn (California: ABC-CLIO, 2014), p. 329.
7 Gyorgy, p. 112.
8 'Nuclear Insurance and Disaster Relief,' *nrc.org* (2014), available online at www.nrc.gov/reading-rm/doc-collections/fact-sheets/nuclear-insurance.pdf (accessed August 8, 2016), p. 1.
9 Wolfgang Saxon, 'G. C. Minor, 62, an Engineer Who Criticized Nuclear Power', *The New York Times* (July 31, 1999), available online at www.nytimes.com/1999/07/31/us/g-c-minor-62-an-engineer-who-criticized-nuclear-power.html (accessed August 1, 2016).
10 Jack Anderson, 'Carter's Nuclear Team', *The Dispatch*, April 19, 1979, p. 4.
11 Noted in Henderson, p. 334.
12 For a detailed list and full timeline see *Bulletin of the Atomic Scientists*, 'Timeline', *thebulletin.org* (2016), available online at http://thebulletin.org/timeline (accessed August 8, 2016).

Preface

Nuclear Culture

The night was well underway and the local city of Harrisburg was long asleep when a series of mechanical malfunctions and human failures led to the worst commercial nuclear power accident in America. On March 28, 1979 the Three Mile Island Unit-2 reactor experienced a partial meltdown. The news of such an unprecedented accident on American soil sparked a media frenzy; conflicting reports alongside anxious rumors contributed to a sense of chaos within the local community but also in America as a whole. The accident became a critical moment in American history and is an integral part of a vibrant and ongoing cultural debate surrounding the development and use of nuclear technology.

Three Mile Island is positioned within a long and fascinating history of rapid technological developments following the influential events of World War II (1939–1945). For some, the iconic sight of the cooling towers stretching high over the Susquehanna River in Pennsylvania continues to provoke fear of destruction and potential apocalypse. The Chernobyl disaster of 1986, and the more recent 2011 Fukushima Daiichi crisis only facilitated an exacerbation of dystopian perspectives. For others, the sight of the active cooling towers today represents the triumph of the American authorities in containing and overcoming the Three Mile Island incident.

What impact did the Three Mile Island incident of 1979 have on the people? How did this event articulate deep-seated tensions rooted in World War II? In what ways did this crisis inform ongoing debate following the Chernobyl catastrophe and the Fukushima disaster? Although *Three Mile Island: The Meltdown Crisis and Nuclear Power in American Popular Culture* addresses these critical questions, this book will primarily deal with the lingering fear, but also ambivalence, towards nuclear technology present in American culture. Three Mile Island was as much a social and political issue as it was a problem of nuclear physics. The focus of this book lingers not in the control room, but on the State of Pennsylvania, the country, and the wider world. This book aims to forge links between the Three Mile Island crisis and its place within the wider public debate on nuclear technology. To do this it is necessary to navigate a range of primary sources, responses in popular culture, and critical scholarly work.[1] Over the following five chapters, we will engage with a range of theorists, authors, and cultural historians, and examine the interconnection between history and imagination, between fact and fiction, and between public reaction and industry response. A cultural-historical reading of mainstream responses in American society—from news stories to science fiction—is one way, of many, to view a genuinely fascinating and influential moment in the nation's past.

In an academic journal article from the 1980s, William A. Gamson and Andre Modigliani discuss nuclear power in media discourse and public opinion and explain that "nuclear power, like every policy issue, has a culture".[2] It is this idea of culture that is under investigation here. To shape and engage with culture is, for Michael Denning who writes extensively on this issue, a process of toil, struggle, and expression as important as any type of work. As Denning puts it, "Culture had become an industry in the early twentieth century, and artists, musicians, and writers were laborers in that industry."[3] The construction of American culture was pivotal to cementing a sense of American identity that was distinguishable from that of the colonial countries; it would also come to help shape understandings of the present and pave the way for the future evolution of societies and groups. When we speak of culture today we might be speaking about large issues in terms of global cultures (as well as Western culture, Eastern culture, African culture, and so on). We could also be speaking about the values (often differing) of a small group within a wider collective and we might refer to this as a subculture. We may think of cultural hierarchies of high culture and low culture; we might even speak of bespoke group characteristics such as 'teen culture'. At the heart of these terms and ideas is an attempt to articulate something about the beliefs, customs, practices, ideals, workings, and processes of a society, group, or even the mood of a time, place, or event. We can, as Gamson and Modigliani note, also have a 'nuclear culture' and this culture can be as complex, nuanced, and diverse as any other. Like clouds in the sky through which we might see faces and images, the mushroom cloud as an icon of the nuclear age can be widely interpreted and narrated.

Popular culture specifically concerns the everyday and mainstream elements that contribute to social experience, such as entertainment; it involves that which uniquely shapes and reflects social values and ideas.

Popular culture can be understood as a reflection of dominant and prevailing *popular* behaviors and beliefs characteristic of a group at a particular time. Popular culture content enables debate on political and social issues through numerous media in a way that helps the masses both navigate and communicate urgent matters facing the community (be they local, national, or worldwide communities). In many ways, popular culture is both an intervention in, and mirroring of, political discourse. Popular culture, rather than offering a simplification of important political arguments, often reflects what official reports and technological studies can fail to address adequately: how mainstream society shapes, and is shaped by, landmark events such as a nuclear accident. When we speak of 'nuclear popular culture' we are talking about how certain nuclear issues and debates are represented through all manner of public facing products from news stories to film. In this book, popular culture will help us navigate the many narratives that seem to have bloomed from the nuclear cloud, a cloud that has loomed over all nuclear developments since August 1945 with the first—and only—use of a nuclear weapon during war. Popular culture is an important medium through which to examine and analyze the impact of America's worst commercial nuclear power incident on both the local community and the nation as a whole.

While the examination of nuclear accidents will naturally position a lot of the presented arguments in a dystopian space, I am neither a critic nor supporter of nuclear power here. I am also not a nuclear physicist who will provide technical insight into Three Mile Island as a pressurized water reactor plant, nor scientific commentary on the accident itself. Chiefly, it is my intent to explore how the public found meaning in (and shaped cultural understandings of) the so-called nuclear age to clarify how events like Three Mile Island affected and shaped how America viewed, and continues to view, nuclear technology. I have tailored this book for the general reader and those seeking an introduction to 'nuclear culture' and twentieth century American technological history within the context of Three Mile Island. The first two chapters introduce the wider historical framing of nuclear development in American culture, including how nuclear technology is part of what one might regard as advancement in genocide technologies. As I will explain, many nuclear tensions prevalent in society today can be traced back to anxieties surrounding information and misinformation which have plagued and problematized portrayals of nuclear technology—and the nuclear industry—since the nuclear bomb was dropped on Hiroshima. Chapter 2 focuses more on Atoms for Peace and nuclear power development; the importance of the Pennsylvanian community and environment around Three Mile Island in the plant's early life will also be considered. While Chapter 3 looks at the Three Mile

Island crisis in terms of its cultural impact in detail, Chapter 4 examines how the event was articulated and discussed in popular culture expressions such as film, literature, television, art, political cartoons, and graphic novels. The final chapter offers an overall assessment of how disasters like Chernobyl and Fukushima marked the endurance of nuclear anxiety and also help to keep the Three Mile Island crisis relevant. I initially wanted call this book 'Nuclear Reactions' but decided on a more descriptive title; nevertheless, this book is, in its heart and soul, about nuclear reactions in all their forms.

NOTES

1 Unlike Harry Henderson's *Nuclear Power*, this text avoids offering a robust history of nuclear power and does not make contact with the hard science of the nuclear industry. While *TMI 25 Years Later: The Three Mile Island Nuclear Power Plant Accident and Its Impact* by Bonnie A. Osif, Anthony J. Baratta, and Thomas W. Conkling claims understanding of nuclear technology is required for a study of Three Mile Island, this is not necessarily the case when exploring popular culture. Here we are asking how the Three Mile Island incident impacted the people. This book is not intended to be a comprehensive survey of nuclear technology in America like Robert C. Williams and Philip L. Cantelon's *The American Atom*. The span of this book cannot be this ambitious. Further, unlike Robert Del Tredici's *The People of Three Mile Island*, I do not concentrate on a focused exploration of first person accounts from the time. Instead of asking *what happened* and *to whom*, this book asks what happened *before* and *after* the incident through the exploration of broad social response.

2 William A. Gamson and Andre Modigliani, 'Media Discourse and Public Opinion on Nuclear Power: A Constructionist Approach', *American Journal of Sociology*, 95. 1 (1989), 1–37 (p. 1).

3 Michael Denning, *The Cultural Front: The Laboring of American Culture in the Twentieth Century* (London: Verso, 1997), p. xvii.

CHAPTER 1

Atoms for War

World War II and the Cultural History of Early Nuclear Development

Popular culture has long been a medium through which to explore moments of trauma. Popular culture was a useful way for the public to navigate World War II, especially the Holocaust and the bombing of Japan, which were considerable rupture moments in the twentieth century. Emerging from this event was the nuclear age and, again, popular culture became a useful way for the public to navigate their complex responses to the ever-changing world. For Alison M. Scott and Christopher D. Geist, atomic bombs have cast a shadow over Americans and this is explicitly witnessed in popular culture.[1] There is an intimate link in popular culture between the nuclear weapon and nuclear power, often amalgamated as 'nuclear technology' with safety issues, ramifications, and historical significance enmeshed in one term.[2] Before looking at nuclear power and the development of the Three Mile Island plant (Chapter 2), it is first important to explore the 'twin' of nuclear power in cultural theory—the nuclear weapon. Scott and Geist make the bold, but not inaccurate claim that "no phenomenon of major significance in modern world culture and history can be fully appreciated and examined without this new dimension".[3] This new dimension refers to the "new social history" that came from the atomic attack on Japan during World War II; this is never more true than when looking at cultural responses to nuclear power.[4]

It has been difficult, if not impossible, to sever the link between nuclear weapons and nuclear power. This link is something Anna Gyorgy comments on in *No Nukes: Everyone's Guide to Nuclear Power* when she questions what exactly we mean by the term 'nukes'. Gyorgy states, "Any building that contains as much radiation as 1,000 or more Hiroshima-type bombs deserves careful watching."[5] Written in 1979, Gyorgy's text highlights the dangers associated with nuclear plants, naming human error, hardware failures, environmental damage, and natural disasters (such as earthquakes at plant sites) as chief issues. Included in her understanding of the word 'nuke' are plants such as the poorly performing Vermont Yankee station. A public information pamphlet from 1982, questioning the very notion of a 'peaceful' use of atomic technology, also begins by highlighting the association between nuclear power and nuclear weapons: "The enormous power locked in the atom was first revealed to the world in 1945. Since then this power has also been harnessed for generating electricity. But like Siamese twins, the military and civil technologies remain linked."[6]

To fully appreciate the early history surrounding nuclear energy culture, we must start with the technological innovations of World War II.[7] As we will come to see in Chapter 2 onwards, many of the anxieties linked to nuclear weaponry from World War II and the Cold War manifested again during the 1979 Three Mile Island crisis, especially with regards to concern over radiation, health impact, and environmental damage. In Chapter 2 we will also see how there was a concentrated drive in the 1950s towards the 'Atom for Peace' which helped see the development of nuclear power plants; however, before we can look at the peaceful atom we first need to examine the atom of war.

WORLD WAR II: THE RELATIONSHIP BETWEEN NUCLEAR TECHNOLOGY, GENOCIDE, AND APOCALYPSE

Technological advancements have typified and dominated the twentieth century as evidenced by the standard ways of defining epochs: machine age, atomic age, and information age. Philosopher Hannah Arendt claims that the drive towards technology is part of the modern world: "For some time now, a great many scientific endeavors have been directed toward making life also 'artificial', toward cutting the last tie through which even man belongs among the children of nature."[8] The development of technology, especially during World War II, had such a profound impact on the human world, as well as the human condition, that Arendt proposes "a reconsideration of the human condition from the vantage point of our newest experiences and our most recent fears".[9] Our more recent 'experiences and fears' involve nuclear development and the many additional advances that are tied to this crucial innovation. Yet, the development of this revolutionary technology did not occur in a vacuum. In this section, we will explore how World War II saw the unprecedented advancement of what we might call 'technologies of genocide' and how the nuclear weapon, acting as a landmark example of a new type of warfare, rose out of a war characterized by industrialized killing. The development of nuclear technology during the last World War links this nuclear innovation to arguably one of the most rupturing and psychologically, economically, and politically traumatic events of recent history.

The Holocaust was a rupture event marking unprecedented mechanized mass slaughter and a new level of dehumanization. The trauma of this event resonated throughout the twentieth century as an extraordinary episode, one that American Holocaust literature scholar Lawrence Langer describes as causing "a permanent hole in the ozone layer of history".[10] The enormous tragedy of this period has caused many people, from academics to eyewitnesses, to attempt to comprehend the events of World War II artistically and critically. One notable example is Primo Levi, an Italian Jewish pharmacist who was transported to Auschwitz in 1944. Levi not only wrote numerous texts on the Holocaust (*Auschwitz Report* (1946), *If This is a Man* (1958), *The Truce* (1963), *The Drowned and the Saved* (1986)), but also many science fiction stories and essays detailing the potentially apocalyptic use of technology. Levi's fiction expresses concern that fearsome technology would be the new threat facing humankind with similar potential for victimization and dehumanization as experienced during the Holocaust. In fact, much of the language used by Levi relating to the Holocaust and Nazi Party concentrates on concerns over how technology and artificial structures furthered Nazi goals; this was something fellow survivor Filip Müller also reflects on when he refers to "extermination machinery" and "the extermination technique".[11] Levi notes that part of the success of the Nazi regime was down to "German technological and organizational perfectionism".[12] Beyond the Nazi party, technology itself was a defining part of World War II and heralded the advent of advanced machines such as the V-2 and the nuclear weapon. Both World Wars saw the production of revolutionary technological innovations that seemed to position technology as instrumental to winning a modern war.

One of the reasons technology played such a vital role in World War II was that it enabled humans to act beyond their bodily restrictions, and this is instrumental when considering the capacity of nuclear technology in warfare because the weapon allows the military to act beyond physical limitations of hand-to-hand combat. Nuclear energy also, as we shall see, enables the power industry to move beyond traditional and finite power means (such as costly and limited fossil fuels). While remote warfare existed in early conflicts through the use of arrows and cannons, weapons like the V-2 and the atomic bomb demonstrated that World War II was characterized by distance weapons in a way no previous war was. This is something Stephan J. Kline, writing in 1985, reflects on when he argues that the construction of weapons (like guns and atom bombs) extends the human ability to kill and defend.[13]

Building on the potential for war technology to extend the capabilities of the human, psychologist Robert J. Lifton notes the benefit of technology as enabling distance between the assailant and the target; this distance seems to be enough to change concepts of mortality, guilt, and emotional resonance.[14] The extensive range and capability of remote warfare to target large areas and groups of people from incredible distances makes "the high technology of destruction compatible with genocide".[15] American philosopher of technology, Lewis Mumford, argues that before the technological progression of World War II genocide was "restricted by the amount of hand labor required".[16] One example of the evolution from murder by hand to large structures of genocide can be witnessed in the shift in tactics by the Einsatzgruppen (Nazi special task force) who initially exterminated target groups by gunshot. It was found that mass shootings were inefficient, and Einsatzgruppen soldiers experienced difficulties committing close range murder; consequently, gas vans and chambers were

The Manhattan Project

The fear of the Nazis developing an atomic weapon led, in part, to the founding of the Manhattan Engineering District (otherwise known as the Manhattan Project) by President Franklin D. Roosevelt. The group, formed in 1942, was tasked with creating an atomic bomb before the enemy. The Manhattan Project was not merely a small group of scientists working on the technology in a small room of a government building. The project had enormous scale with a staff of over 130,000 and funding that amounted to almost $2 billion. The Manhattan Project was spread over many sites including Canada and the United Kingdom.

The first successful demonstration of a controlled nuclear chain reaction was the Chicago Pile 1 in December 1942. The Chicago Pile was the first nuclear reactor and was named after the location (The University of Chicago) and the structure of the experiment (a pile described by Manhattan Project physicist and Nobel Prize winner Enrico Fermi as a pile of bricks and timbers layered with uranium). Alongside the development of the nuclear weapon, the 'pile' also demonstrated that the reactor could produce heat which could be used to generate electricity. However, at this time, the focus was on how to use the science for military purposes through the construction of a nuclear weapon—nuclear energy for civilian purposes was a linked, but shelved potential.

implemented. Killing at a distance was found to minimize, or ease, the psychological distress experienced by killers. Mumford notes that distance weapons, like the atom bomb, enable psychological protection in a similar way.[17]

The ultimate example of nuclear potential was revealed to the world when The Manhattan Project culminated on August 6, 1945, after the Enola Gay bomber dropped the uranium nuclear bomb dubbed 'Little Boy' on the Japanese city of Hiroshima. The Bock's Car (also known as Bockscar) bomber unleashed the plutonium weapon 'Fat Man' on the city of Nagasaki three days later. On August 6, after the detonation of Little Boy, President Harry S. Truman announced the atomic bombing of Hiroshima to the American people and the wider world; in the first part of his speech he spoke of the enormous importance the bomb had in winning the war:

THE TRINITY TEST

The Trinity Test took place on July 16, 1945 in New Mexico and was overseen by Robert J. Oppenheimer, the scientific director of the Manhattan Project. As Brigadier General Thomas F. Farrell noted after watching the Trinity Test, "No man-made phenomenon of such tremendous power had ever occurred before."[18]

> Sixteen hours ago an American airplane dropped one bomb on Hiroshima and destroyed its usefulness to the enemy. That bomb had more power than 20,000 tons of TNT. It had more than two thousand times the blast power of the British "Grand Slam" which is the largest bomb ever yet used in the history of warfare.[19]

Although the nuclear bombing of Japan was accredited with shortening World War II and reducing military casualties (as outlined by President Truman), the devastating ramifications of the weapon created an instant and long-lasting psychological impact. Captain Robert A. Lewis, the co-pilot of the Enola Gay, wrote, "My God, what have we done?" on the journey back to America. The effects of August 6, 1945 were catastrophic: "a searing fireball", "a giant mushroom cloud", and "a fire-storm" that "uniformly and extensively devastated" the cities.[20] Blast and fallout impacted population, environment, and economy and there were cases of social collapse and severe psychological side-effects. Subsequent ramifications included radiation-induced death and severe health conditions.[21]

It became apparent that nuclear weapons represented an "unprecedented peril" in which usage "could end human civilization".[22] Arguably, the ability for *one* bomb to cause mass annihilation on an unprecedented scale presented technology as acting to "'modernize' the genocidal system"[23] and became central to a modern understanding of genocidal structures. For many, the nuclear weapon came to represent the ability to "destroy the world".[24] In H. D. Smyth's *A General Account of the Development of Methods of Using Atomic Energy for Military Purposes under the Auspices of the United States Government 1940–1945* the atomic bomb is described as "potentially destructive beyond the wildest nightmares of the imagination".[25] Unlike previous weapons like the V-2, the atomic bomb would come not only to define World War II but to define, in many respects, the century; as Lifton remarks, "You cannot understand the twentieth century without Hiroshima."[26] Although the Holocaust and the atomic bombings of the Japanese cities were radically different in execution, intent, and ideology, Levi links the Holocaust and Hiroshima together as landmark trauma events which define the twentieth century.[27] And, if World War II saw the escalation of genocide structures, the nuclear weapon was seen as representing the potential for the ultimate genocide act—extinction. This is why the nuclear threat and the atrocities of the Holocaust are not separate for Levi. Levi articulates metaphorical and literal links between the two events: "The exodus of brains from Germany and Italy, together with the fear of being surpassed by Nazi scientists, gave birth to the nuclear bombs."[28] It is not merely the case in Levi's thought that one event provoked the other, but rather that extreme violence begat extreme violence. Nuclear technology became an explicit example of how catastrophic technological advancements could be. As Victor Brombert explains of Levi's sentiment, "Not only could the Nazi nightmare be repeated . . . but modern science, unleashing nuclear forces, threatened to put an end to all human life."[29]

The nuclear threat continued through the Cold War. The significance of the nuclear attack in 1945 endures even today, prompting Andrew J. Rotter to state: "More than sixty years after the bombings of Hiroshima and Nagasaki, people still have nuclear nightmares."[30] This is perhaps why, in his later works such as *The Drowned and the Saved*, Levi identifies nuclear technology as one of the dangers that face the youth of today: "These young people are besieged by today's problems, different, urgent: the nuclear threat."[31] Thinkers also following this sentiment, such as Arendt, suggest that nuclear technology is so revolutionary that the modern world "was born with the first atomic explosions".[32] F. G. Gosling notes that the attack on Hiroshima was deeply strategic, designed to "make a profound psychological impression on the Japanese and weaken military resistance"; it also had a profound effect on the American public.[33] Although the bomb was initially met positively as a saving force of the war, it was after witnessing the catastrophic events of the bomb that trauma emerged: "When the atomic bomb was dropped over Hiroshima,

Americans felt both deep satisfaction and deep anxiety, and these responses have coexisted ever since."[34]

Alongside many anxieties that emerged following the realization of the atomic bomb's devastating ramifications, the public was eventually exposed to photographs and reports of the weapon's 'inhumanity'. Although previous weapons had caused severe bodily and psychological trauma, depictions and witness reports of the effects of radiation and blast damage on the human form were shocking—especially as the attack on Hiroshima involved just one weapon. After Hiroshima, the nuclear bomb seeped into every avenue of American existence and resonated as a dominant part of American thought. Peggy Rosenthal explains that the mushroom cloud is now a symbol of the atomic age and is enshrined with cultural meaning (from politics to history and from triumph to disaster).[35] Frequently, the mushroom cloud is associated with radiation and radiation became a term that would linger over all nuclear development.

THE COLD WAR

The Cold War (origins in 1945 and conclusion in 1991) principally involved the United States of America and the Soviet Union as they struggled for power and dominance in the post-World War II world. Although a nuclear weapon volley seemed close in October 1962, the Cold War was termed as such because the United States and Soviet Union never actually engaged in battle, leaving the war 'cold' rather than 'hot'.

Secular Nuclear Apocalypse

Apocalypse is perhaps an apt term to describe attitudes towards how some viewed the potential of nuclear technology in the immediate World War II period. Although historically the term 'apocalypse' is linked to biblical meaning, today, the word apocalypse has come to have a more secular sense regarding widespread disaster, often alluding to "immense cataclysm or destruction".[36] Roslyn Weaver extends apocalypse to cover "widespread destruction of land or the urban environment" through scenarios such as war, biological attacks, and nuclear technology.[37] In fact, Robert J. Lifton explains that ideas of nuclear extinction revolutionized apocalypse and through the nuclear weapon a new apocalyptic vision has emerged.[38] So not only do we see a shift from Holocaust to Hiroshima regarding technologized killing, but we also see at this time a firm shift away from religious to secular ideas of an apocalypse. An apocalypse caused by man, not God.

Figure 1.1 Mushroom cloud
Department of Defense

NUCLEAR EDUCATION: LEARNING ABOUT THE NEW BOMB AND NEW TECHNOLOGY

On one day the world was presented with a whole new science and with it came new terminology, new debates, and an entirely new vision of the future.[39] It would be a while before new generations were taught the basic principles in school and longer still before history classes were able to analyze the 1945 event. Simply put, while the scientific industry had been preparing for the atomic bomb for years, the general public woke up to a new bomb and a new future. The question of how to respond to this sudden and rupturing development left many in a state of sheer shock and ambivalence. What is nuclear technology? How does it work? What are the benefits and dangers? What really happened in Japan? What does the nuclear future look like? Questions like these needed answers. As a result, many publications (from film to cartoons) attempted to explain this new technology to the public.

One very early example that attempted to respond to the nuclear bombing of Japan was the film *The Beginning or the End* (1947), directed by Norman Taurog. The film premiered in Washington D.C. and attempted to explain and justify the use of nuclear weapons at the end of World War II. Taurog's film relates the development and utilization of the nuclear weapon from the days of the Manhattan Project to the deployment of the weapon over Hiroshima. However, critics largely dismissed the film as government propaganda meant to suggest to the American public that the bombing of the Japanese city was crucial. Many scientists after the 1945 detonation engaged in the nuclear debate and also attempted to educate the public on the new technology. A flurry of publications including *Atomic Energy in the Coming Era* (1945), *Our Atomic World* (1946), *Our World or None* (1946), and *The Atomic Story* (1949) sought to explain the science and question the future of the technology.[40] The edited volume *Our World or None* includes contributions from the likes of Hans Bethe, Albert Einstein, J. Robert Oppenheimer, and Leo Szilard. The editors Dexter Masters and Katharine Way explain that the rationale behind the book was for it to assist citizens in understanding the new 'challenge' presented by atomic technology.[41] The difficulty in comprehending this astonishing new force led to movements by both pro- and anti-nuclear groups to explain the technology in clear and straightforward ways. Pocket Books attempted to explain and explore the atomic bomb in a collection called *The Atomic Age Opens*, published in the same month as the nuclear bombings of Japan.

Cartoons and comics after 1945 also responded to the rupturing nuclear moment. Many comic books fused sensationalism with the fantastic; it was in the 1940s that Superman went head to head against Atom Man. There was also a rush of atomic animal stories, as Ferenc Morton Szasz notes, aimed at introducing atomic humor after Hiroshima.[42] However, sensationalism aside, many comic books also attempted more seriously to provide a nuclear education for the layperson struggling to understand the new technology. In the first edition of *Science Comics* (January 1946), the comics presented "wonders of science in pictures"; the first story was

'The Bomb that Won the War'.[43] The cover image features the distant city of Hiroshima smothered by a mushroom cloud with a plane in the foreground. The start of the story reads: "On August 5, 1945, a single bomb from a single B-29 devastated the Japanese city of Hiroshima—and with its detonation the world entered a new era: The Atomic Age!"[44] Trying to convey the scientific aspect of the development, *Science Comics* uses white marbles to represent electrons and yellow marbles to represent the atom's nucleus in order to portray 'atom smashing'. This educational, storytelling device was clearly aimed to impress the relevant science upon readers in a way that would be easy to comprehend and enjoy. Amidst the story of the bomb, several frames, strategically intermingled with scenes of bomb development and explosions, reflect on the potential for peaceful atomic energy. The comic concludes by offering reassurance that America is in control of this new science, and peaceful use can be made of the war-time technology. *True Comics* (March 1946) ran a similar story entitled 'Atoms Unleashed: The Story of the Atomic Bomb'. *True Comics* concentrated on light retellings of non-fiction stories. Like *Science Comics*, this too features several frames that explain the splitting of atom. Again, the final panels (once more juxtaposed with artwork depicting the Hiroshima bombing) concentrate on peaceful uses of the technology. The 'peaceful atom' and the Atoms for Peace project will be considered in the next chapter.

Already, we can see that following the apocalyptic experience of the bomb there came a drive to allay public fears. However, many publications also drew attention to some potential dangers of the technology. In *Picture News* comics (January 1946) the cover asks "Will the Atom Bomb Blow

Figure 1.2 'Atoms Unleashed: The Story of the Atomic Bomb'
True Comics, 47 (New York: March 1946)

the World Apart?"[45] The comic book, aimed at families, addresses this question and while the article issues the reassuring line, "Don't be frightened."[46] *Picture News* does speak of the need for caution: "Like a Sorcerer's Apprentice, our magic, without knowing how to stop itself, will destroy us, if we don't watch out."[47] The voice of the story is George Bernard Shaw, speaking in the first person about the wonders, benefits, and also perils of atomic power. A nuclear education is also provided by *Future World Comics* (1946) in a piece called 'How Atomic Energy Works', in which a laundry tub is used to describe the movement of an atom, a balloon used to define critical mass, and a coal pile used to illustrate a chain reaction.[48] However, *Future World Comics* also presents the dangers of atomic power in the hands of mad scientists specifically. In this story there is a more dystopian take on nuclear technology, and a concentrated shift from fears over nuclear weapons to fears over nuclear power with both being described as having explosive potential. In this story, the publication deals with a nuclear plant under threat with the exclamation that disaster is imminent: "If I don't get that switch closed the whole plant may blow up!"[49] Fortunately, the hero Bill Cosmo defeats the mad scientist and saves the day. This story also marks a change in concerns from fears over a nuclear weapon being deliberately used for assault to a new worry over nuclear power being negligently controlled.

Along the same vein as educational comics, some political cartoons also sought to explain the political climate surrounding nuclear development as well as issue warnings and attempt to balance fact and fiction. Political cartoons are a time-honored and established practice in America stretching back to May 9, 1754 (*Join, or Die* cartoon published in the *Pennsylvania Gazette* by Benjamin Franklin) and play a large part in shaping both how America discusses politics and engages with social issues. Consequently, when nuclear power entered public experience, political cartooning responded to the new development. Cartoonist Herbert Block (known as Herblock) was a popular, trusted, and consistent political artist in the press and over the eight decades of Block's influence, the nuclear theme featured in a range of his cartoons. Some of Herblock's cartoons present a tall personified nuclear bomb character, often with the word 'Atom' scrawled somewhere on his body.[50] The illustrations in which the character 'Atom' often feature see the character loom large over small, oblivious or anxious politicians; in a few cartoons, 'Atom' is portrayed as bigger than the globe and overlooks planet Earth like a predator. Often 'Atom' is seen with a clock or makes reference to ticking, as if to say time is short in the nuclear age and like the Doomsday Clock (*Bulletin of the Atomic Scientists*) warns, the fate of the human race is close to midnight and apocalypse. Herblock's cartoons demonstrate the evolution of nuclear concerns—moving from concerns over Hiroshima and tensions between the Soviet Union and the United States towards later issues of global nuclear terrorism.[51] Herblock's cartoons first featured in the *Washington Post* in 1946, but over the following decades 'Atom' was never far away.

Throughout these examples, there is a concentrated effort to explain a sudden and unpresented technology. While some publications attempt to explain the hard science, others act to highlight the potential perils. Even in educational pieces like *Science Comics*, the mushroom cloud and apocalyptic taint is always in the background. The Cold War, as we shall see, only added to lingering apocalyptic concerns.

"UNTIL THE RUSSIANS MADE ONE TOO . . .": THE COLD WAR AND NUCLEAR DEVELOPMENT

Nuclear education coincided with the Cold War (1945–1991), and as the people started to learn more about this paradoxically fearsome and wondrous technology, they also had to contend with a new conflict. In many respects, the 1950s was oversaturated with problems but the struggle between the Soviet Union and the United States of America was filled with intense paranoia, suspicion, and animosity. Distrust of the Soviet Union was deeply embedded in American culture; the Soviets' censorship, The People's Commissariat for Internal Affairs (NKVD), and Communist dictatorship appeared to fundamentally conflict with America's belief in freedom, capitalism, and democracy. Also, Communism prevented the private ownership of land and the private means to produce and self-sustain as celebrated under American capitalism. Furthermore, the Soviet people were viewed as suppressed under Stalin's dictatorship (1924–1953), particularly by the NKVD (1934–1954) who focused on monitoring the people, locating enemies, and suppressing dissenters.[52] Not only was the censorship particularly excessive, but the government propaganda was also known for being "unprecedentedly intense".[53] Even after the death of Joseph Stalin in 1953 and the attempt at De-Stalinization by Nikita Khrushchev, concern over the Soviet Union and the rise of Communism still dominated American thought during the Cold War. In short, the Soviet worldview was seen to challenge the American perspective of what it meant to lead a moral and free human existence.[54]

Historians accuse American Senator Joseph R. McCarthy of intentionally exploiting the "atmosphere of fear and suspicion" during the 1950s through his claims that the communist threat was everywhere and had infiltrated State Department Officials.[55] It was during the paranoia of the McCarthy era that Klaus Fuchs was found guilty of sharing secrets of the nuclear project with the Soviet Union (1950), and the married couple, Ethel and Julius Rosenberg, was accused of being communist spies who leaked vital intelligence on the atomic bomb to the Soviets. If anything would ignite growing Cold War tensions, it was nuclear technology. The villainization of the Soviet Union contributed to growing alarm when they were found to be developing and testing their own atomic weapons. Not only was America confronted with a communist menace and a potential new war but it was faced with the terrifying realization that a regime that politically and ethically opposed American ideals may also perfect nuclear weaponry.

Before the Soviets developed nuclear weapons, the United States often found triumph, delight, and humor in the A-bomb. The popularity and cultural relevance of the atomic bomb made it a buzzword important in marketing and saw its use in a range of scenarios including the naming of a cocktail in the Washington Press Club bar.[56] The atom dominated radio, Hollywood, journalism, music, novels, and poetry. American children were also viewed as an audience for selling 'the atomic spectacular' as many toys and advertisements were aimed at the young with the popularization of atomic board games, collectibles, laboratory kits, fashion accessories, books, magazines, and comics.

The merge of science with advertising and merchandise is not a huge surprise as, before the bomb, scientific advancements often played important roles in product conceptualization. The word 'radium' was used to suggest a product was a luxury and a valuable commodity in the early part of the twentieth century; there was a range of 'radium brand' butter, cigarettes, alcohol, matches, and even condoms. The X-ray brand also produced a range of polishes, soap, and razor blades. Although it was not unusual for companies to use scientific buzzwords in products for the mass market, the atom became a central point in many facets of human existence including everyday products, recreation, medicine, leisure, and even swimwear. "Radiation and radioactivity provoked interest of an intensity unprecedented in American history for a scientific topic" and this interest predated the twentieth century.[57]

The origins lay in the late nineteenth century (primarily 1896 with the X-ray) and caused considerable awe and interest for the public. Some early accounts featured talk of marvels and miracles; for example, at the turn of the twentieth century a 'cure' for a young blind girl called Tillie Spitznadel was accredited in the press to the 'restoring' effects of "remarkable" radium.[58] The magical nature of radium especially was referred to in a lecture by dancer Loïe Fuller in 1911 who, inspired by the physics, incorporated her enchantment with science into her performances. Magic, for Fuller, was the best way to describe radium for "magic is a mystery and we call a thing a mystery because we do not understand it".[59] The magic Fuller imagined was the potential for radium to bring forth amazing discoveries. Fuller even wondered if radium would one day be able to photograph the soul leaving a dying body. So great was the interest in new 'magical' scientific developments (like the X-ray and discoveries like radiation) that Matthew Lavine refers to them as crazes.[60] However, after 1945, Paul Boyer attributes positive nuclear attitudes to the winning of World War II and the defeat of an 'inhuman' enemy.[61] Thus, the atom was more than just an exciting piece of jargon and was instead representative of an impressive advancement; it represented the winning of the war, the vanquishing perhaps of 'inhumanity' itself. As such, some who did experience guilt, fear, trepidation, or uncertainty over America's nuclear power may have chosen silence due to perceiving their cynicism as somewhat unpatriotic. The nuclear weapon 'won the war', so to speak openly of it being anything other than a saving tool was potentially a social taboo and may not have been well received.

Imagination ran rife as the many advantages of atomic energy were envisioned. Most of these dreams were ambitious and suggested utopian situations such as cheap and endless energy. However, after the atomic bomb Joe-1 was tested by the Soviet Union in August 1949, America suddenly had to think about nuclear shelters and radiation—all previously abstract and irrelevant issues. Tony Hilfer explains that the jubilant mood of the American population after World War II changed radically once Joe-1 was detonated: "the atom was initially perceived as a scientific marvel, a form of white magic—until the Russians made one too".[62] The atomic bomb was no longer perceived as under the control of the United States government. The threat was now close to home and looming ever closer: "Almost overnight the assumed permanency of life on earth had vanished, and people were forced to live with the traumatic awareness that total, worldwide obliteration was a strong possibility in the near future."[63] The difference in how the nuclear weapon was viewed when in the hands of America compared to how it was depicted in the hands of the Soviets was highlighted by political cartoonist Pat Oliphant in his 1983 inked cartoon called 'Rockets'. Here, the Soviets' rocket is described with negative terms likening the Soviets with evil, while America's rocket is linked to positive terms and presented as on the side of God.[64]

Nuclear testing by both sides wrenched nuclear imagination out of World War II and repositioned it in the Cold War setting. Nuclear testing became a constant psychological stress for many Americans who feared both the tests and the 'inevitable' war for which the tests seemed to be preparing. Lifton explains that initially the nuclear threat appeared to be abstractly distant and that the Japanese appeared to be "guinea pigs" for the newest weapon; however, the testing in Nevada during the 1950s prompted Americans to declare that they too were nuclear "guinea pigs".[65] By 1951, the United States had conducted twenty-four nuclear tests; but they were not alone, for the Soviet Union had also started atomic weapon testing and after their first detonation tested a further three. It was during the 1950s that discussions were held over whether the United States should pursue the super weapon, a thermonuclear device (otherwise known as the hydrogen bomb). Oppenheimer noted that no scientist was prepared to endorse the further development of thermonuclear weaponry due to the limitless capabilities of the weapon and its potential to be used for genocide and stated that "a super bomb should never be produced".[66] Encouraged by Lewis Strauss and Edward Teller (both leading and influential figures in nuclear weapon development), President Truman directed that work on the hydrogen bomb should continue, a decision nuclear physicist Hans Bethe called a "calamity".[67] Thus, in 1952, Ivy Mike became the first thermonuclear bomb to be tested; this test was followed by Soviet Union tests one year later.

A TIME OF AMBIVALENCE

Previously, we saw how many publications after 1945 sought to educate the American people on what exactly nuclear technology was. We saw how many publications seemed torn between presenting factual science and issuing warning messages; we also saw how some publications expressed enthusiasm for the 'unlimited potential' of nuclear technology. We also saw how, before the Soviets' development of Joe-1, the atom seemed fun and entertaining for Americans enjoying post-war buoyancy. During the Cold War Arms Race, in the face of Soviet nuclear development, we also saw a duality of experience in which nuclear technology was viewed as both magnificent and terrifying. The meshing of polar responses—from fear to awe—and the interpretation of nuclear technology as both apocalyptic and amazing led to a strong vein of nuclear ambivalence in American culture. The nuclear weapon was not only a scientific marvel but the explosion was a spectacle, a thing of beauty and the United States had mastery over it. However, the Soviet Union had it too and a nuclear World War III, which could cause human extinction, seemed possible. Consequently, public opinion really did oscillate—the bomb was seen as both a friend and a foe.

David Nye, in his work on the technological sublime (which refers to the magnificence of geographical, industrial, or architectural structures which conjure strong emotional responses like fascination, pleasure, or pain) also reports this sense of ambivalence. Nye notes that the first atomic test rendered the scientists awed, triumphant, and humbled.[68] Such opposed reactions had been recorded by Brigadier General Thomas F. Farrell, who witnessed the Trinity Test and described the internal conflict the sight caused: "The effects could well be called unprecedented, magnificent, beautiful, stupendous and terrifying."[69] During the 1950s, the power of the atom bomb also became splendid for the tourists who would travel to witness atomic tests. Nye notes how the Nevada desert was visited by schoolchildren and families hoping to witness an atomic detonation at the test site.[70] Upon seeing the explosion and resulting mushroom cloud, witnesses would experience a mixture of wonder at American scientific achievement but also fear that this technology could turn on them—especially now it was in the hands of the Soviet Union.

Fondness for the atomic spectacle, as seen through the practice of nuclear tourism, was also notably represented in souvenir photographs and readily available in postcard form. Soon after Hiroshima and Nagasaki, photo postcards were created and distributed marking the bombings; many were educational. In the 1950s, postcards showing nuclear events, such as nuclear tests, were popular, so popular that whole books have been compiled on them. Atomic postcards were purchased to keep as mementos, some were also gifted to others, and many were preserved and even displayed in frames. On the back of an *Atomic Explosion: Frenchman's Flats, or Yucca Flats, Nevada* postcard, a personal message reads "I know that you will want to have this pictured framed".[71] Exactly why such images were preserved and framed is something collection editors John O'Brian and Jeremy Borsos question when they ponder if the picture is framed "as a source of pleasure, or . . . as a source of caution?"[72]

While some postcards were clearly cautionary and part of a mourning act (such as *The Dead Man's Shadow*, which is the photograph of a human shadow burnt into steps in Hiroshima, c. 1950), others were directly used as a source of pleasure. In late August 1945, postcards advertising a showing of the "atomic bomb explosion" were mailed out. The advertisement by Embassy Newsreel Theatre (for New York and Pennsylvania) seemed to celebrate the nuclear event by referring to the end of the war, using uppercase and exclamation marks to promote excitement: "SEE! FOR THE FIRST TIME THE DEVASTATING FORCE OF THE NEW BOMB THAT BROUGHT ABRUPT END TO PACIFIC WAR." The card highlights the 'selling points' of the event—the newness of the technology, the extraordinary power, and the link between America's new weapon and the conclusion of the war.[73] But, the card also mentions devastation. On the back of a postcard named *Spectacular Nuclear Explosion* (a United States Army image from Enewetak Atoll), a handwritten note reads "This was practically right in our front yard (last July) Quite a site to watch (sic)".[74] The homophone 'site' works well to refer to the moment as a *place* to watch and also as a *thing* to watch; seemingly unconcerned by the proximity, the sender of this postcard articulates what Nye suggests about the pleasure of watching nuclear explosions. Yet, pivotally, many photo postcards of the time did not just glorify in the spectacular mushroom cloud image. Many postcards were of towns (such as Oak Ridge, Tennessee, and White Rock Los Alamos County, New Mexico) celebrated as being built and having thrived in response to nuclear development in these areas. Such postcards praised the jobs created and the successful communities established. The nuclear power plant for energy production would also come to be linked to (and lauded for) creating jobs, settling families in a region, and contributing to large community development. On a mushroom cloud postcard by the Chamber of Commerce Farmington, New Mexico, a little packet of uranium ore is accompanied with the positive claims of the advantages available in Farmington (such as work on drilling rigs, constant sunshine, and oil, gas, and uranium industries).[75] Clearly trying to 'advertise' the area, the postcard aligns itself primarily with the nuclear to encourage interest; the nuclear explosion is certainly positive here as "Like an 'Atomic Bomb'", the city is described as "Bustin'-Out".[76]

Literature of the time also recorded conflicting perspectives. Many science fiction texts attempted to present what they considered to be plausible nuclear scenarios, yet many had an apocalyptic feel. For scholar Paul Brians, who has written extensively on nuclear issues in fiction, apocalyptic 'realism' is essential in all nuclear fiction texts. Brians, commenting on nuclear fiction for children, argues that any nuclear fiction text which enshrines themes of hope is "unhelpful" and grossly misleading.[77] For Brians, then, apocalypse is linked to plausibility: nuclear stories have to be dystopian if they are to be 'realistic'. Many stories seem to follow this concept. Ray Bradbury's *There Will Come Soft Rains* (1950) depicts (implied) extinction following a nuclear war. Many of Philip Wylie's texts address the catastrophe of nuclear war: *Blunder* (1946) and *Tomorrow!* (1954) are notable examples of texts which feature extensive atomic devastation. Both Leigh Brackett's *The Long Tomorrow* (1955) and Philip K. Dick's *The World Jones Made* (1956) feature post-apocalyptic America. In addition, Nevil Shute's post-apocalyptic World War II novel *On the Beach* (1957) received critical praise for portraying a 'plausible' nuclear scenario. All these texts feature likely scenarios and repercussions such as death, radiation, dehumanization, the collapse of civilization, and destruction. However, perhaps in an attempt to counter the sheer volume of sensationalist atomic fiction, many fiction magazines included factual essays on nuclear technology. *Astounding Science Fiction* magazine, for example, published 'Destruction from Atomic Weapons' by Lt. Theodore S. Simpson (1954) which presented facts about nuclear weapons; 'Tornadoes and Atom Blasts' by J. O. Hutton (1954) argued that severe weather could be linked to atomic testing; and 'On Atomic Jets' by J. J. Coupling (1955) focused on the potentials of atomic energy.

Film also played a major role in representing nuclear technology and again nuclear technology stories seemed to be split between the sensationalist and insouciant movie. There was also an effort to defuse nuclear fear and soothe tensions. During World War II, the American government recognized the ability of Hollywood to affect public opinion, and so the Office of War Information (OWI) reviewed and influenced scripts.[78] In conjunction with the Office of Censorship, the OWI educated the filmmaker on what should be contained in the film and what should be left out.[79] After 1945, the government's influence and censorship extended to Hollywood productions which featured any nuclear content.[80] Due to government restraints, many early Hollywood depictions of the nuclear theme either befriended the nuclear bomb or used satire and fantasy to defuse concerns (such as *The Beast from 20,000 Fathoms*, 1953, in which a dinosaur is revived through atomic testing).[81] Many Hollywood texts of the time veered away from 'realism' and utilized the buzzwords 'nuclear', 'atomic', 'atom', and 'radiation' to fuel exaggeration and sensationalized stories of superheroes/villains, mutation, and alien worlds. The idea seemed to be that if the threat was made to seem silly and unrealistic, then anxieties would be diffused. Yet, even if plots in *The Incredible Shrinking Man* (1957) and *Attack of the Crab Monsters* (1957) were deemed too ridiculous to be taken as anything other than over-the-top science fiction, the fear that nuclear contamination could lead to hellish mutation and danger was very real. Although the plot of the 1954 science fiction film *Them!* is outlandish because of its monstrous irradiated bugs, the message it carries is rather relatable: "When man entered the atomic age, he opened a door into a new world. What we eventually find in that new world, nobody can predict."[82] Mutant crustaceans merely act as metaphors for how horrific, transformative, and frightening nuclear contamination could be, even though such concerns are delivered through comic exaggeration. In films like *The Incredible Shrinking Man*, *Attack of the Crab Monsters*, and *Them!*, radiation has the power to change humans and the natural world into something fundamentally alien, unnatural, and inhuman. Michael L. Lewis argues that science fiction about nuclear technology during the Cold War covered both the bomb as a "doomsday machine" and the bomb as "bringing out world peace".[83] While this is true, the 1950s fiction also oscillated between portraying the bomb as horror as well as comedy; many examples presented stories oscillating between realism and sensationalism. Widely different approaches highlighted not only the ambivalence of the public towards nuclear technology during the 1950s, but they also illustrated the struggle in popular culture to make sense of a new technology and a changing world.

Music of the time evidenced this tension by presenting the bomb as a fun spectacle or apocalyptic symbol, representing general attitudes towards nuclear technology and reflecting many different debate points through lyrics. Plenty of songs approached the atomic age in a light-hearted way and sought to cash in on the nuclear craze. Slim Gaillard's lively and jovial *Atomic Cocktail* from 1946 celebrates the dawn of the atomic age with an easy jazz sound. Many songs like the Chords' hit *Sh'Boom* (1954), although seemingly inspired by the bomb, do not offer melody or lyrics that have any relevance to the technology. The only reasonable link made between the 'sh'boom' of an atomic explosion and the lyrics is a vague assumption that romance can be explosive. Similarly, *Atom Bomb Baby* by The Five Stars (1957) compares the attractiveness of a woman to the nuclear bomb; like the bomb, the female subject of the song is more explosive and 'hotter' than TNT. *Atomic Love* by Little Caesar and the Red Callender Sextette (1953) describes love as atomic, and Chris Cerf's *Fallout Filly* (1961) uses nuclear imagery to speak of a temperamental woman. Such songs show the cultural relevance of anything nuclear but with little understanding of how to adequately or appropriately utilize it. The nuclear had social importance; people wanted to write about it, talk about it, and even sing about it, but could not always communicate about it successfully. In many cases, to be culturally relevant and to engage with a popular moment within mass culture, a link was seen to be needed to the nuclear—even if it did not make much sense.

However, other songs explored the dangers and ramifications of the nuclear with serious intent. *When That Hell Bomb Falls* by Fred Kirby (1950) describes

> A weapon of destruction
> To destroy us everyone.[84]

Old Man Atom in 1950 by The Sons of the Pioneers explores atomic history and offers warnings about the atom's frightful reality.[85] In an early example from 1946, *Atomic Power* by The Buchanan Brothers reflect on the bombings in Japan and warn that if nuclear power is misused America could be on the receiving end.[86] The song also makes early reference to the potential peaceful uses of the new technology which should be pursued instead: "use it for the good of man and never to destroy".[87]

While many songs reflected fears, some tunes attempted to soothe with rousing camaraderie. Hank Williams' *No, No, Joe* (1950) acts as an open letter to Joseph Stalin but also to the American public about American resilience and supremacy. Here, strength is not only encouraged through being a good and tough American but through a renewed focus on Christianity. While songs like *No, No, Joe* talk of the weapon being a Godly power and therefore a gift, other songs speak of religious faith as essential during the Atomic Age. In 1946, *When the Atom Bomb Fell* by Karl and Harty speaks of the bomb as an act of God. *Atomic Telephone* by Spirit of Memphis Quartet (1952) also links God to atomic power, as does The Louvin Brothers' *Great Atomic Power* (1952) when they talk of the "Savior in the air".[88] In *Jesus Hits Like an Atom Bomb* (1951), the group combines nuclear concerns with the gospel, encouraging the people to take the Lord as seriously as they take nuclear war. In 1959, Tom Lehrer entertained television audiences with his satirical 'survival song' *We'll All Go Together When We Go*, which comically talks of apocalypse and extinction:

> For if the bomb that drops on you
> Gets your friends and neighbors too
> There'll be nobody left behind to grieve
> And we will all go together when we go.[89]

Much laughter was heard during the performance as fallout, funerals, bomb, and ash, along with references to people baking, frying, burning, and dying were placated by a clever rhyme and lively tune; the song proved itself essential to defusing tension through a very frank and nonchalant play on cultural concerns. The song almost seems to suggest that there is no point in worrying about the bomb because the fate will be shared and instantaneous. In fact, it seems silly to worry at all. Here, soothing through encouraging listeners not to give in to fear (often implied as a weakness and even un-American), and to put faith in the wisdom of God helped to reassure and allay fears. It will all be ok, and if it is not, then we will 'all go together' anyway.

SOOTHING ACTS

In the 1950s, America was in a strange position as it continued to rejoice in the winning of the war while faced with a whole range of other conflicts that presented the decade as a time of unrest and pending crisis. Civil Defense was a public focused campaign with advertisements appearing everywhere from milk cartons to huge billboards; President John F. Kennedy even wrote an open letter on the importance of homemade nuclear shelters, published in *Life* (1961).[90] While Civil Defense was a massive movement with a reported 50,000 advertisements, exhibits, and educational materials published, few Americans fully embraced Civil Defense in their homes (few installing nuclear shelters).[91]

A DAY CALLED 'X'

The CBS mockumentary *A Day Called 'X'* (1957), set in Portland, Oregon, shows the community evacuating amidst fears of a Soviet nuclear attack. *A Day Called X* shows and encourages calmness and rationality in the face of nuclear attack and evacuation.

Perhaps the lack of action was a result of the efforts of the media and government to assuage the burgeoning fear of the American people during the 1950s. As Boyer explains, the government attempted a "soothing" of atomic fears which led to acceptance and even an appreciation of the nuclear.[92] The film *The Medical Aspects of Nuclear Radiation* (c. 1950) lightens the fear of radiation by willfully misleading the public through unrealistic scenarios focused on pointing out the many types of radiation that are safe. In regards to the nuclear bomb, *The Medical Aspects of Nuclear Radiation* flippantly suggests "be somewhere else when it happens" and if this is not possible to simply protect against it.[93] The film notes that if contamination does occur treatment is available, and ultimately concludes that devoting "eighty-five percent of one's worrying capacity" to radiation is a "fallacy" and "unsound".[94] *The Medical Aspects of Nuclear Radiation* encourages audiences to be afraid of what they can see—blast damage and fires (which are themselves familiar wartime occurrences). *Fallout* (1955) puts such a positive spin on a nuclear attack that post-disaster shelter life is compared to a holiday in which the adults are advised to bring favorite drinks, tinned foods, books, and toys for children—"some of the same things you might take on a vacation camping trip".[95] *Fallout* even reminds families to take food seasoning like salt, pepper, and sugar for culinary comfort, dramatically lessening the severity of the potential nuclear crisis. *Survival Under Atomic Attack* (1951) shows misleading footage of partially destroyed Japanese buildings alongside tranquil scenes of healthy and happy Japanese families; the inference is that the Hiroshima civilians were relatively unaffected by the atomic bomb: "the majority of people exposed to radiation recovered completely, including a large percentage of those who suffered serious radiation sickness. Today they lead normal lives."[96] The film *Duck and Cover* featuring Bert the Turtle (1951) subliminally acts to show that even the frailest and slowest of animals will survive the bomb.

Comics of the time, as expert Ferenc Morton Szasz explains, often presented atomic technology as a beneficial force for the future, with publications like *The Atomic Age* overlooking radiation and Hiroshima; Szasz argues books like these from the 1950s "reflected the mood of the Cold War era".[97] Many corporations started to employ literature as a medium through which to educate and sell the public on new science, namely atomic science. Thus, as Szasz notes, companies like General Electric produced their own comics to sell children on atomic power.[98] Leonard Rifas, a prolific anti-nuclear comic book contributor, explains that comic books sponsored by utilities (such as General Electric's 1948 *Adventures Inside the Atom*) factually presented nuclear power as safe, limitless, and wondrous. This, combined with awesome nuclear-enhanced superpowers, would have influenced young readers to view the technology as "strength and power".[99] Such bias by the utilities provoked Rifas to produce *All-Atomic Comics* to offer the other side of the nuclear argument—the anti-nuclear stance.[100] Most industry and government propaganda encouraged the audience to be 'calm' through the guarantee that survivability was practically assured: somewhere and somehow an American hero would prevent disaster. The placatory scenarios offered by the media and Hollywood acted to subvert prevalent concerns over the bomb by presenting it as familiar and relatively safe. This did not just soothe tensions over nuclear weapons and war; such an understanding would be pivotal to ensuring compliance when nuclear reactors would be positioned near towns and cities. The message, across the board, needed to be that nuclear power, in all forms, was safe and beneficial.

While there were many examples of nuclear 'soothing' at the time, one of the most successful acts of 'soothing' came via Walt Disney's publication and corresponding television documentary *Our Friend the Atom* (1957). Walt Disney was extremely influential at the time as both a source of education and entertainment. Disney had managed to penetrate popular culture with over eighteen films, 145 shorts and twenty documentaries. Furthermore, Disney played a significant role in the lives of Americans and delivered many educational texts aimed at children and adults; one example is *Cleanliness Brings Health* (1945). Critics celebrated and supported Disney's *Our Friend the Atom* for presenting the bomb positively within popular culture and "easing stress regarding this powerful force in the modern world".[101] *Our Friend the Atom* was a small educational segment of a Disney documentary series discussing nuclear technology. The documentary series operated through genres known as 'Frontierland', 'Futureland', 'Adventureland', and 'Fantasyland'. 'Futureland' was marketed as the realm of the atom and the "the promise of things to come".[102] Disney was the perpetual comforter of the day, using a heavy dose of hallucinogenic fantasy to belay facts and the medium of gentle narration and innocent cartoons to educate. A. Bowdoin Van Riper notes that it was Disney's duty to bring cheer to a technologically "ambivalent society" during an era of post, present, and prospective trauma.[103] However, the American people were not wholly ambivalent; Scott Bukatman notes that many feared the future: "No longer was 'The Future' a harmless fiction, a utopian era that, by its very definition, will never arrive; it was, instead, upon us, and with a vengeance."[104] Disney's documentary was an act of soothing by transforming the atom from fiend to friend.

The Disney documentary heavily relies on the narrative device of *The Fisherman and the Genie* to explain the advent of the nuclear age. This tale is an amalgamation of *The Story of the Fisherman* and *The Story of the Young King of the Black Isles* from *The Arabian Nights* collection. In *The Arabian Nights*, the narrator is a young bride called Scheherazade who, to delay the Sultan's next murder, distracts the Sultan with the story of a fisherman who finds a golden pot inhabited by a powerful genie. At first, the genie is dangerous and threatens to kill the fisherman but quickly the fisherman outwits the creature and manages to control him. In the Disney version, the genie is a Nuclear Genie and the golden pot is uranium. The symbolism is clear: humankind will control the Nuclear Genie and reap the rewards. Disney points out the beneficial effects of atomic research on advancing the human race but at no time explicitly refers to danger. For example, in the film Marie Curie's role in the story of atomic development is told. Curie poetically states that she will call her discovery radium but, in Disney's retelling, her radiation fueled death is omitted from the story. More importantly, the words Hiroshima, Nagasaki, and Japan are never mentioned. Van Riper states that Disney's dealings with the Bomb "sidestepped, glossed over, or altogether ignored the consequence".[105] In the final frame, the narrator tells the audience that if we remember the spirit in which the technology was intended then "the atom will become truly our friend".[106] The friendly aspect of nuclear technology, the audience is led to believe, is nuclear energy. A nuclear power plant, fueled by the Nuclear Genie, can light cities and grow crops: a peaceful atom to soothe years of fear and turmoil.

CONCLUSION: A MOVE TOWARDS PEACE

The development of nuclear technology in the twentieth century has experienced peaks and troughs with regards to American public opinion. In this introductory chapter, we have navigated some of the many ways nuclear technology was viewed and how the wider context of World War

II and the Cold War would have helped to shape public experience of nuclear technology. We have explored how nuclear technology was developed during a period of intense technological and industrial killing, and how it was linked to an evolution of mass killing structures. Through looking at thinkers such as Primo Levi, we have seen how anxieties over nuclear technology were intimately linked to both the Holocaust and to Hiroshima. We have also noted the longevity of such anxieties as the nuclear race between the principle Cold War nations contributed to unrest surrounding the future of nuclear technology and the potential for a nuclear World War III. However, we have also understood how there was an attempt to soothe nuclear fears with companies such as Walt Disney redirecting public focus to the benefits of nuclear energy in an effort to shed the darker connotations of the technology.

Already we can see the early beginnings of nuclear energy development through this history. From the Chicago Pile 1 in 1942 to Disney's Nuclear Genie in 1957, we can see the roots of nuclear energy starting to take shape during a chaotic time marked by a fair amount of skepticism and fear on one hand and sheer ambivalence on the other. The 1950s is an apt decade in which to break away from our examination of the nuclear weapon, for the 1950s saw the establishment of the world's first nuclear power stations, as we shall see in the next chapter.

In Chapter 2, many of the themes uncovered here will resurface. The Cold War Arms Race continued and increasing tensions between the United States and Soviet Union, as well as spectacular and destructive nuclear tests, ensured that the violent atom continued to have dominance in the public experience of nuclear technology. However, through the establishment and development of the nuclear power industry, the atom was not only positioned as a 'friend' (in Disney's words) but due to the deliberate severance in rhetoric between nuclear war and nuclear power, there was a concentrated effort to present nuclear energy as the peaceful alternative to its weaponized form. Nuclear power as a peaceful friend also appealed to the 'magical' quality that seemed so enchanting for the likes of Loïe Fuller in the early twentieth century. Nuclear weapons and war seemed linked more explicitly to apocalypse and a Godly sense of power and were therefore more fearsome and destructive; whereas the ability to power cities alongside additional dreams (like nuclear cars) seemed to position nuclear power as more exciting and beneficial for the American people. In many respects, nuclear power enabled a psychological break from nuclear trauma through repositioning the technology in peaceful contexts.

However, large nuclear accidents called into question this polarization between the dangerous weapon and the peaceful plant. Further, as we shall see in the next chapter, 'peace' projects like Plowshare would also complicate the direct split between violent atom and peaceful atom through the use of 'peacetime' explosions. Accidents at plants, which were reported in the media as posing radiation risks to workers and local communities, would also confuse the idea that plants were entirely without threat. Of course, when things go wrong at a nuclear power plant and terms like 'radiation' are cited as a risk, then the swirling nebula of nuclear context that includes war, Hiroshima, explosions, fallout, sickness, and danger are suddenly applied (to various extents) to the industry that was claimed to be peaceful. In Chapter 3, the Three Mile Island crisis will be examined in detail; however, before this point it is important to examine how nuclear power was presented as the peaceful and friendly twin of the nuclear weapon—a twin that could be relied on, trusted, and embraced. Nuclear power accidents—like Windscale (1957), Three Mile Island (1979), Chernobyl (1986), and Fukushima (2011)—would express nuclear 'violence' in a new way. It is partly due to the past efforts of soothing, and the presentation of the industry as magical, peaceful, friendly, and benevolent, that such crises became as rupturing and disturbing as the war context from which the power was born.

NOTES

1 Alison M. Scott and Christopher D. Geist, 'Preface', in *The Writing on the Cloud: American Culture Confronts the Atomic Bomb*, ed. by Alison M. Scott and Christopher D. Geist (Oxford: University Press of America, 1997), pp. v–1 (p. v).
2 As this book stretches from 1945-2017, the terms 'atomic' and 'nuclear' are used variably in a range of sources. Therefore, while there is technically a difference in meaning between these terms, this book, like public discourse and other texts exploring nuclear culture, prefers interchangeable usage.
3 Alison M. Scott and Christopher D. Geist, pp. v–1 (p. v).
4 Alison M. Scott and Christopher D. Geist, pp. v–1 (p. xi).
5 Anna Gyorgy, *No Nukes: Everyone's Guide to Nuclear Power* (Montreal: Black Rose Books, 1979), p. 99.
6 *Atoms for Peace?* (Commonwork Land Trust Information Service on Energy, 1982), p. 1.
7 Matthew Lavine, who speaks of atomic ages (the first occurring for him between 1895–1945; the second after 1945), argues that to start with World War II and talk of anxieties towards uranium and radium is to reflect, or even engage in, "cultural amnesia". For Lavine, a true timeline starts around 1895 and 1896 with the discovery of X-rays and radiation. However, this book is focused on the context surrounding popular cultural reactions to Three Mile Island and as such starting this discussion in the late nineteenth century is neither practical nor helpful. Matthew Lavine, *The First Atomic Age. Scientists, Radiations, and the American Public 1895–1945* (New York: Palgrave Macmillan, 2013), p. 3.

8 Hannah Arendt, *The Human Condition*, trans. by Margaret Canovan, 2nd edn (London: University of Chicago Press, 1998), pp. 2–3.
9 Arendt, *The Human Condition*, p. 6.
10 Lawrence L. Langer, *Holocaust Testimonies: The Ruins of Memory* (London: Yale University Press, 1991), p. xv.
11 Filip Müller, *Eyewitness Auschwitz*, ed. by Susanne Flatauer (Chicago: Ivan R. Dee, 1999), p. 160.
12 Primo Levi, *The Drowned and the Saved*, trans. by Raymond Rosenthal (London: Joseph, 1986), p. 66.
13 Stephan J. Kline, 'What is Technology', *Bulletin of Science, Technology & Society*, 1.215 (1985), 215–218 (p. 216).
14 Robert J. Lifton, *The Nazi Doctors. Medical Killing and the Psychology of Genocide* (New York: Basic Books, 2000), p. 495.
15 Lifton, *The Nazi Doctors*, p. 495.
16 Lewis Mumford, *The Myth of the Machine* (London: Secker and Warburg, 1971), pp. 180–181.
17 Mumford, *The Myth of the Machine*, p. 267.
18 Quoted in Martin J. Sherwin, *A World Destroyed: The Atomic Bomb and the Grand Alliance* (New York, 1977), p. 312.
19 Harry S. Truman, 'Statement by the President of the United States', Washington D.C., The White House, August 6, 1945, in *Harry S. Truman Library & Museum* (2016), available online at www.trumanlibrary.org/whistlestop/study_collections/bomb/large/documents/index.php?documentid=59&pagenumber=1 (accessed July 19, 2016).
20 Michael J. Hogan, 'Hiroshima in History and Memory: An Introduction', in *Hiroshima in History and Memory* (Cambridge: Cambridge University Press, 1999), pp. 1–10 (p. 1); *United States Strategic Bombing Survey: The Effects of the Atomic Bombings of Hiroshima and Nagasaki* (June 19, 1946), pp. 1–51, in *Harry S. Truman Library & Museum* (2016), available online at www.trumanlibrary.org/whistlestop/study_collections/bomb/large/documents/index.php?documentdate=1946–06–19&documentid=65&studycollectionid=abomb&pagenumber=1 (accessed July 5, 2016); The Committee for the Complication of Materials of Damage Caused by the Atomic Bomb in Hiroshima and Nagasaki, *Hiroshima and Nagasaki: The Physical, Medical, Social Effects of the Atomic Bombings*, trans. by Eisei Ishikawa and David L. Swain (London: Hutchinson, 1981).
21 Note that it took a long time before the American public was made aware of the consequences of the Japanese blasts—such as realizing the full details of the suffering. Therefore, initial focus was on the might of the bomb and the splendor of the explosion, what Joyce Nelson would refer to as a "radical separation of the bomb and the human body" (Joyce Nelson, *The Perfect Machine* (Ontario: Between the Lines, 1987), p. 33).
22 Robert J. Lifton and Eric Markusen, *The Genocide Mentality* (New York: Basic Books, 1990), pp. 2, 3.
23 Lifton and Markusen, *The Genocide Mentality*, pp. 2, 3.
24 Lifton, *The Nazi Doctors*, p. 495.
25 H. D. Smyth, *A General Account of the Development of Methods of Using Atomic Energy for Military Purposes under the Auspices of the United States Government 1940–1945* (Washington, D.C., 1945), p. 163.

26 Robert J. Lifton and Greg Mitchell, *Hiroshima in America, Fifty Years of Denial* (New York: A Grosset/Putnam Book, 1995), p. xi.
27 Anthony Rudolf, 'Primo Levi in London', in *The Voice of Memory*, by Primo Levi, ed. by Marco Belpoliti and Robert Gordon (Cambridge: Polity Press, 2001), pp. 23–33 (p. 29).
28 Levi, *The Drowned and the Saved*, p. 168.
29 Victor Brombert, *Musings on Mortality: From Tolstoy to Primo Levi* (London: The University of Chicago Press, 2013), p. 160.
30 Andrew J. Rotter, *Hiroshima: The World's Bomb* (Oxford: Oxford University Press, 2009), p. 304.
31 Levi, *The Drowned and the Saved*, p. 166.
32 Arendt, *The Human Condition*, p. 6.
33 F. G. Gosling and The United States Department of Energy, *The Manhattan Project: Making the Atomic Bomb* (Washington: Department of Energy, 1999), p. 46.
34 Lifton and Mitchell, *Hiroshima in America*, p. xi.
35 Peggy Rosenthal, 'The Nuclear Mushroom Cloud as Cultural Image', *American Literary History*, 3.1 (1991), 63–92.
36 John Wallis, 'Apocalypse at the Millennium', in *The End All Around Us: Apocalyptic Texts and Popular Culture*, ed. by John Wallis and Kenneth G. C. Newport (London: Equinox Publishing, 2009), pp. 71–97 (p. 73).
37 Roslyn Weaver, 'The Shadow of the End. The Appeal of Apocalypse in Literary Science Fiction', in *The End All Around Us: Apocalyptic Texts and Popular Culture*, ed. by John Wallis and Kenneth G. C. Newport (London: Equinox Publishing, 2009), pp. 173–198 (p. 174).
38 Robert J. Lifton, 'NUCLEAR', *New York Times*, September 26, 1982, p. 58.
39 Ferenc Morton Szasz's *Atomic Comics* highlights that the public were aware of atomic breakthroughs before 1945. Coverage from the 1930s to the 1940s is beyond the parameters of this book. Ferenc Morton Szasz, *Atomic Comics* (Las Vegas: University of Nevada Press, 2012).
40 Szasz, *Atomic Comics*, p. 43.
41 *Our World or None*, ed. by Dexter Masters and Katharine Way (Whittlesey House, 1946).
42 Szasz, *Atomic Comics*, p. 63.
43 *Science Comics*, 1, (Springfield: Humor Publications, January 1946), p. cover.
44 'The Bomb that Won the War', *Science Comics*, 1, (Springfield: Humor Publications, January 1946), p. 1.
45 *Picture News*, 1.1. (New York: January 1946), p. cover.
46 'Could Science Blow the World Apart?' *Picture News*, 1.1. (New York: January 1946), p. 3. N.B. the title of the story differs from the front cover which is phrased: 'Will the Atom Bomb Blow the World Apart?'.
47 'Could Science Blow the World Apart?', p. 6.
48 'How Atomic Energy Works', *Future World Comics*, 1 (New York: Summer 1946), p. 41.
49 'Bill Cosmo and the Plutonium Pile', *Future World Comics*, 1 (New York: Summer 1946), p. 39.
50 See "Pardon Me, Mister—Do You Know What Time It Is?" (*Washington Post*, 1946), "Don't Mind Me—Just Go Right On Talking" (*Washington Post*, 1947), and "Tick-Tock Tick-Tock" (*Washington Post*, 1949).

51 See untitled image of Quaddafi, Saddam Hussein, Assad, Khomeini dreaming of nuclear bombs from 1984.
52 Sarah Davies, *Popular Opinion in Stalin's Russia: Terror, Propaganda, and Dissent, 1934–1941* (Cambridge: Cambridge University Press, 1997), p. 14.
53 Davies, *Popular Opinion in Stalin's Russia*, p. 4.
54 The fear of Communism was witnessed before the Cold War during the Red Scare (1919–1920).
55 Ronald J. Oakley, *God's Country: America in the Fifties* (New York: Dembner Books, 1986), p. 56.
56 Paul Boyer, *By the Bomb's Early Light: American Thought and Culture at the Atomic Age* (New York: The University of North Carolina Press, 1994), p. 10.
57 Lavine, *The First Atomic Age*, p. 25.
58 See Lavine, *The First Atomic Age*, Chapter 2.
59 Loïe Fuller, 'Lecture on Radium' (London, January 20, 1911).
60 Lavine, *The First Atomic Age*, Chapter 2.
61 Boyer, *By the Bomb's Early Light*, pp. 12, 24.
62 Tony Hilfer, *American Fiction Since 1940* (Harlow: Longman, 1992), p. 4.
63 John Brosnan, *Future Tense. The Cinema of Science Fiction* (London: Macdonald and Jane's Publishers, 1978), p. 72.
64 Pat Oliphant, 'Rockets', in *The New World Order in Drawing and Sculpture* (Missouri: Universal Press Syndicate Company, 1994), p. 16.
65 Robert, J. Lifton, 'On the Nuclear Altar', *New York Times*, July 26, 1979, p. 19.
66 'USAEC General Advisory Committee Report on the "Super", October 30, 1949', in *The American Atom. A Documentary History of Nuclear Policies from the Discovery of Fission to the Present. 1939–1984*, ed. by Robert Williams and Philip Cantelon (Philadelphia: University of Pennsylvania Press, 1984), pp. 120–128 (pp. 123–127).
67 Hans Bethe, 'Hans Bethe, Comments on the History of the H-Bomb', in *The American Atom. A Documentary History of Nuclear Policies from the Discovery of Fission to the Present. 1939–1984*, ed. by Williams and Cantelon, pp. 132–141 (p. 140).
68 David Nye, *American Technological Sublime* (Massachusetts: MIT, 1994), p. 228.
69 Alwyn McKay, *The Making of the Atomic Age* (Oxford: Oxford University Press, 1984), p. 100.
70 Nye, *American Technological Sublime*, p. 233.
71 From Friod and Family, in John O'Brian and Jeremy Borsos, *Atomic Postcards* (Chicago: Intellect, 2011), p. 9.
72 O'Brian and Borsos, *Atomic Postcards*, p. 10.
73 O'Brian and Borsos, *Atomic Postcards*, p. 22.
74 O'Brian and Borsos, *Atomic Postcards*, p. 36.
75 'Bustin'-Out! Like An Atomic Bomb', in O'Brian and Borsos, *Atomic Postcards*, p. 71.
76 O'Brian and Borsos, *Atomic Postcards*, p. 71.
77 Paul Brians, 'Nuclear War Fiction for Young Readers: A Commentary and Annotated Bibliography', in *Science Fiction, Social Conflict and War*, ed. by John Philip Davies (Manchester: Manchester University Press, 1990), pp. 132–151 (p. 133).

78 Clayton R. Koppes and Gregory D. Black, *Hollywood Goes to War. How Politics, Profits and Propaganda Shaped World War Two Movies* (Los Angeles: University of California Press, 1990), p. vii.
79 Koppes and Black, *Hollywood Goes to War*, p. 324.
80 There remained a lack of focus on Hiroshima and Nagasaki in film throughout the 1940s and 1950s; yet there were many films about nuclear catastrophe. There seemed to be a psychological divide in which people did not want to face the reality of the Japanese bombings but had a need to express some of their concerns in fiction. The *Beginning or the End* needed presidential authorization and the numerous factual issues with the film marked for Lifton "The Hollywoodization of the bomb" (Lifton and Mitchell, *Hiroshima in America*, p. 361).
81 There were some dystopian examples such as *Rocketship XM* (1950), *Five* (1951) and *Unknown World* (1951).
82 *Them!*, dir. by Gordon Douglas (Warner Bros, 1954).
83 Michael L. Lewis, 'From Science to Science Fiction: Leo Szilard and Fictional Persuasion', in *The Writing on the Cloud: American Culture Confronts the Atomic Bomb*, ed. by Scott and Geist, pp. 95–105 (p. 95).
84 Fred Kirby, *When That Hell Bomb Falls* (Atomic Platters: Cold War Music from the Golden Age of Homeland Security, 2010).
85 The Sons of the Pioneers, *Old Man Atom* (Atomic Platters: Cold War Music from the Golden Age of Homeland Security, 2010).
86 The Buchanan Brothers, *Atomic Power* (Atomic Platters: Cold War Music from the Golden Age of Homeland Security, 2010).
87 The Buchanan Brothers, *Atomic Power.*
88 The Louvin Brothers, *Great Atomic Power* (Atomic Platters: Cold War Music from the Golden Age of Homeland Security, 2010).
89 Tom Lehrer, *We'll All Go Together When We Go* (Lehrer Records, 1959).
90 J. F. Kennedy, 'A Message to You from the President', *Life*, September 15, 1961, p. 95.
91 John Gregory Stocke, '"Suicide on the Instalment Plan:" Cold-War-Era Civil Defense and Consumerism in the United States', in *The Writing on the Cloud: American Culture Confronts the Atomic Bomb*, ed. by Scott and Geist, pp. 45–60 (p. 46).
92 Boyer, *By the Bomb's Early Light*, pp. 303, 334.
93 *The Medical Aspects of Nuclear Radiation*, U.S. Air Force (Cascade Pictures of California, c. 1950).
94 *The Medical Aspects of Nuclear Radiation.*
95 *Fallout*, Office of Civil and Defense Mobilization (Creative Arts Studio, 1955).
96 *Survival Under Atomic Attack*, Official United States Civil Defense Film (Castle Films, 1951).
97 Szasz, *Atomic Comics*, p. 69.
98 Szasz, *Atomic Comics*, p. 70.
99 Leonard Rifas, 'Cartooning and Nuclear Power: From Industry Advertising to Activist Uprising and Beyond', *PS: Political Science and Politics*, 40.2 (April, 2007), 255–260 (p. 255).
100 Rifas, 'Cartooning and Nuclear Power', pp. 255–260 (p. 256).

101 Steven Watts, *The Magic Kingdom: Walt Disney and the American Way of Life* (Missouri: University Missouri Press, 1997), p. 313.
102 *Our Friend the Atom*, dir. by Hamilton Luske (Walt Disney Productions, 1957).
103 A. Bowdoin Van Riper, 'The Promise of Things to Come', in *Learning from Mickey, Donald and Walt. Essays on Disney's Entertainment Films*, ed. by A. Bowdoin Van Riper (North Carolina: McFarland, 2011), pp. 84–103 (p. 85).
104 Scott Bukatman, 'There's Always Tomorrowland: Disney and the Hypercinematic Experience', *Reviewed Works*, 57 (1991), 55–78 (p. 59).
105 Van Riper, 'The Promise of Things to Come', pp. 84–103 (p. 95).
106 *Our Friend the Atom.*

CHAPTER 2

Atoms for Peace

Nuclear Power and the Influence of the Long 1960s

In 1939, Albert Einstein wrote to President F. D. Roosevelt about the research Enrico Fermi and Leo Szilard had been conducting and announced: "the element uranium may be turned into a new and important source of energy in the immediate future".[1] Einstein's first concern in this letter was the delivery of a message about nuclear energy; however, in the third paragraph Einstein presented the sobering fact that this technology "would also lead to the construction of bombs".[2] On that point, Einstein was convinced; the resolute word 'would' revealed his certainty. In the conclusion of his letter, Einstein delivered worrying news that "Germany has actually stopped the sale of uranium from the Czechoslovakian mines which she has taken over"—the implication being that Germany was working on the very same technology.[3] Initially, the need for nuclear bombs was not only more pressing than nuclear power but more noteworthy as well, as commented on in the 1945 'Smyth Report' by physicist H. D. Smyth: "The expected military advances of uranium bombs were far more spectacular than those of a uranium power plant."[4]

Now, in this chapter, we return to Einstein's original vision of "a new and important source of energy" as we shift from anxieties over nuclear weaponry and destruction to the 'domestication' and civilization of the technology. Here, we will examine various social changes and technological advancements in America during the period of the Three Mile Island plant's development. Broadly speaking, this chapter contends with what might be called the 'long 1960s' which, for the purposes of this discussion, can be dated as falling between 1957 and 1974. The year of 1957 saw the world's first major nuclear power accident in the United Kingdom at Windscale, and 1974 marks the completed construction of Three Mile

Island's Unit-1. Unit-1 entered construction in 1967 and Unit-2 began construction in 1970. After this long build-up, after ten years of hope and anxiety, Unit-2 operated commercially for only three months before it experienced crisis and partial meltdown on March 28, 1979. From construction to completion, Three Mile Island saw over a decade of change. In this chapter, we look at some influential developments in both nuclear contexts (such as Atoms for Peace, Windscale, and the Karen Silkwood scandal) and wider historical frameworks (including government projects like Plowshare, local crises like Centralia, and the environmental movement) in order to evaluate in Chapter 3 how such issues may have contributed to (or compounded) how the Three Mile Island 1979 crisis was eventually reported, responded to, and discussed.

Initially, it is important to look at how, against the backdrop of nuclear testing and the Cold War, nuclear technology shifted from a technology of war to a peaceful innovation. How the atom became domestically viable through a fascinating mix of magic and peace rhetoric will be discussed. To do this, we must first examine how the Atoms for Peace project, which originated in the 1950s, established the atom as 'civilized' for the major industry growth that would come during the long 1960s. In the previous chapter, nuclear power was linked to genocide and apocalypse; however, there was another side to the coin and this involved more than just 'soothing'—it was about repositioning the atom as peaceful for use in America's backyard.

FROM NUCLEAR WAR TO NUCLEAR ENERGY: ATOMS FOR PEACE

When Disney, in *Our Friend the Atom* (book, 1956, televised 1957), spoke of the atom becoming our 'friend', Disney was referring implicitly to the Atomic Energy Act of 1954 and the Atoms for Peace project. The Atomic Energy Act declared that "Atomic energy is capable of application for peaceful as well as military purposes".[5] Consequently, the act outlined policies about the development of atomic energy in the interests of general welfare and private enterprise. One of the main purposes of the act was to engender a focused development nationally and internationally for the application of atomic energy for defense, security, and in the interests of public health and safety.

Atoms for Peace was one way to counter the informational brutality of Operation Candor which was a drive in the 1950s to "inform the public of the realities of the 'Age of Peril' ".[6] Following the Jackson Committee report, which noted that the American people did not fully appreciate the perilous times of the Cold War and nuclear age, President Eisenhower was encouraged to address the people and convey to them through radio and televised talks the dangers of Communism, the atomic power held by the Soviet Union, and the many threats facing America. The need for 'candor' is perhaps unsurprising considering the strong vein of nuclear ambivalence noted in the last chapter. However, Robert C. Williams and Philip L. Cantelon note that Eisenhower was concerned about terrifying the public.[7] His fears were well founded. In an article by Stewart Alsop entitled 'Eisenhower Pushes Operation Candor' (September 21, 1953), Alsop spoke of the uncomfortable time of "eye-opening revelations" and declared that the 1950s were a "race for simple survival" with the Soviet Union and the United States equally armed with nuclear weapons.[8] Candor made way to cynicism and pessimism for some; for others, the use of such frankness confirmed what they had assumed: the time was indeed perilous. Atoms for Peace was one way for Eisenhower to deflect attention in a more positive direction and calm the "Age of Peril". As Williams and Cantelon note, "The Atoms for Peace proposal had offered hope to a fearful world."[9]

Although the 'Atoms for Peace' program was launched by President Eisenhower on December 8, 1953, at the 470th Plenary Meeting of the United Nations General Assembly, the idea of the peaceful atom had been

President Truman: Statement by the President Announcing the Use of the A-Bomb at Hiroshima

Atomic energy may in the future supplement the power that now comes from coal, oil, and falling water, but at present it cannot be produced on a basis to compete with them commercially. Before that comes there must be a long period of intensive research. It has never been the habit of the scientists of this country or the policy of this government to withhold from the world scientific knowledge. Normally, therefore, everything about the work with atomic energy would be made public.

But under the present circumstances it is not intended to divulge the technical processes of production or all the military applications. Pending further examination of possible methods of protecting us and the rest of the world from the danger of sudden destruction.

I shall recommend that the Congress of the United States consider promptly the establishment of an appropriate commission to control the production and use of atomic power within the United States. I shall give further consideration and make further recommendations to the Congress as to how atomic power can become a powerful and forceful influence towards the maintenance of world peace.[10]

discussed publically for a while. In fact, President Truman concluded his announcement of the atomic bombing of Hiroshima by highlighting other uses of atomic power and the ability to utilize the technology for global peace.

Truman attempted to emphasize the future benefits of nuclear technology, especially with regards to nuclear energy, by offering hope: not only had atomic technology won the war but in the future it could help solve some of the pressing energy problems the American people were facing. Truman's reassurance that he would personally ensure atomic power was safe, and the reiteration that the technology was focused towards world peace, acted to calm any unrest initially experienced at the start of the speech with the world's first announcement of a new weapon of mass destruction.

Published shortly after the Japanese bombing, the book *The Atomic Age Opens*, released in November that year, also articulates the potential peaceful use of the atom:

> Atomic fission holds great promise for sweeping development by which our civilization may be enriched when peace comes, but the overriding necessities of war have precluded the full exploration of peace-time application of this new knowledge. With the evidence presently at hand, however, it appears inevitable that many useful contributions to the well-being of mankind will ultimately flow from these discoveries when the world situation makes it possible for science and industry to concentrate on these aspects.[11]

The Smyth Report, an official account of wartime nuclear development, further notes the potential future of nuclear development with reference both to weapons and energy, with the latter representing movement towards "the paths of peace". While Smyth was careful to limit expectations by acknowledging that nuclear powered cars were unlikely, he did state that:

> there is a good probability that nuclear power for special purposes could be developed within ten years and that plentiful supplies of radioactive materials can have a profound effect on scientific research and perhaps on the treatment of certain diseases in a similar period.[12]

During this time, as we saw in the last chapter, comics, literature, and film were attempting to explain the science and politics of the bomb to the public. Although, many comic books educated readers on the atomic bomb, in many examples the peaceful potential of atoms was also presented. In *Science Comics*, 1 (1946) attention was given to the amazing and beneficial advances the atom could bring: "It will be possible to propel a ship around the world by smashing the atoms in that one glass of water."[13] Here, as in many comics of the time, it is heavily implied that incredible achievements could be secured with minimal effort through atomic technology; for example, a ship could be powered over extraordinary distances through just a few atoms. In 1950, at the Chicago State Fair, Westinghouse Electric Corporation hosted the popular 'Theatre of the Atom' to hundreds of visitors a day; the event advertised "The Promise for the Future is this New 'Power For Peace' ". The exhibition focused on presenting atomic technology for the layperson and deliberately concentrated on the peaceful atom and avoided reference to atomic warfare.

So, when the Atoms for Peace program was launched by Eisenhower years later, the concept of the peaceful atom was not new; but, the atom still lived in the shadow of the mushroom cloud (not helped by atomic testing on American soil). During Eisenhower's United Nations address, he spoke about Cold War tensions and the arms race. Although Eisenhower initially detailed the military strength of America granted by nuclear weapons, he progressed to articulate a need for peaceful uses of the technology to both enrich diplomatic links and to reduce atomic fear through the pursuit of benevolent and beneficial atomic uses: "The United States knows that if the fearful trend of atomic military build-up can be reversed, this greatest of destructive forces can be developed into a great boon, for the benefit of all mankind."[14] For Eisenhower, Atoms for Peace would help to mark a shift from "inertia imposed by fear" towards "positive progress towards peace".[15] The concept proved to be extremely popular and influential, revolutionizing how atomic energy was being used around the world. Atoms for Peace fed into the United Nations International Conference on the Peaceful Uses of Atomic Energy and the importance of Atoms for Peace was reflected publically by the introduction of an international award for those who had substantially contributed to the movement. Atoms for Peace also appeared on a 1955 blue 3-cent stamp. During the 1950s, films like *Atoms for Peace* (Encyclopaedia Britannica Films, c. 1950), *The Magic of the Atom* (by Monroe Manning, c. 1950), *Atomic Energy as a Force for Good* (1955), *Living with the Atom* (Moody Institute of Science, 1967), *Nuclear Power in the United States* (ERDA, 1971), and *To Develop Peaceful Applications for Nuclear Explosives* (ERDA, 1971), showed pro-nuclear bias.

An early film example, *A is for Atom* (1952), persuades audiences that the peaceful atom is benevolent and useful. Like Disney's *Our Friend the*

Atom, which speaks of a Nuclear Genie, *A is for Atom* speaks of atomic giants. These giants are presented on screen as animated glowing blue superheroes, but these atomic giants are "within man's power, subject to his commands".[16] The short promotional film was developed by General Electric, and the use of cartoons and heroes to portray industry was standard practice. Since 1926, Reddy Kilowatt, a smiling cartoon humanoid with a body formed of red electricity, has been a brand character representing electricity in America. The genies and giants of Disney's and General Electric's imaginations join many cartoons that attempt to put an innocent (even wholesome) smile on a technological industry. But, while Reddy Kilowatt is an electricity based *man*, nuclear power is represented by *giants* and *genies*—powerful creatures of potentially godlike powers, supernatural entities who would trump the measly looking Reddy.

The benevolent 'magic' of nuclear energy was an enduring image used tactically by Atoms for Peace. We might remember from the previous chapter that radium and the atom were initially presented and perceived as magical, mysterious, and phenomenal, featured in a range of products, and were linked to miracles; Atoms for Peace would continue to position the nuclear as a miraculous wonder. It was easier with Atoms for Peace to tightly link nuclear technology to the marvel and enchantment of magic than it was with the devastating bomb and fearsome awe of the mushroom cloud. The connection between the atom and magic is overtly presented during the short film *It's Electric!* In this promotional film, President Eisenhower holds what he refers to as a "neutron rod" (a large stick with a glowing, bulbous head with similarities to a magic wand).[17] Like a wizard, he controls the "awesome power" and "miracle" of the atom.[18] Throughout *It's Electric!* the atom is described through its "peaceful" potential and ability to realize "prayers and dreams". The atom is also a source of "light" (literal and metaphorical) for "civilian use".[19] *It's Electric!* specifically mentions the atom's shift from nuclear holocaust towards civilian use and it is the nuclear wand, handled by the President, that controls it.

Joyce Nelson refers to the President's wand waving as positively promoting the nuclear through "nuclear-pseudo events" and this wand waving, which was part of President Eisenhower's 'enchanted' inauguration of Shippingport construction (September 6, 1954), draws attention to the "magical act".[20] The opening of Shippingport was a major moment for America, and the world; Shippingport in Pennsylvania was the first large-scale nuclear power plant to produce commercial power purely for *peaceful* purposes. The plant was a massive step forward for the nuclear power industry and it was opened with a fantastical display that transformed the President into a magician. The event, as Nelson rightly points out, was contrived; the wand obviously did not serve any purpose other than to equate the atom and Shippingport with a magical process. The spectacle was purely for visual reselling—for pictures, newspapers, and television; the fact the faux wand was wielded by the President gave the moment a degree of sincerity even though the entire set-up was blatantly absurd. Arguably, because nuclear technology was so tightly linked to the awesome spectacle of the mushroom cloud (that as we remember attracted droves of test tourists), a new exciting and visual moment was required

Shippingport Atomic Power Station

The Shippingport Atomic Power Station in Shippingport, Pennsylvania was positioned along the Ohio River. The plant, as the United States Atomic Energy Commission (AEC) stated, served two purposes: "to further peaceful atomic power knowledge and to demonstrate that atoms can help supply our growing electrical needs".[21] On Labor Day, 1954, President Dwight D. Eisenhower used what he called a "neutron rod" to remotely start construction. The President used this rod again to start the reactor, positioning the entire development of Shippingport as a magical and miraculous process. Shippingport came online in 1957 and was the first completely commercial large-scale nuclear power plant "exclusively devoted to peaceful purposes".[22]

In *It's Electric!* the inauguration of Shippingport is shown. The title of this short film repositions the atom not as a nuclear, radiation filled danger, but as electricity itself. As a concept, electricity was familiar and safe to the American people, so describing Shippingport primarily as electric rather than nuclear would have gone some way to ease tensions. Nuclear technology might be a mystery but electricity was used every day in the home and all Shippingport promised to do was to create electricity. Reddy Kilowatt has helped to personify electricity and the electric industry as fun, safe, and innocent since the early 1920s. With Kilowatt and *It's Electric!*, the atom once linked to a potential nuclear war is repositioned as simple electricity, which is not only understandable and desirable but also sellable. Additional documentaries such as *Power and Promise: The Story of Shippingport* reinforce the magic and miracle of the atom and the familiar output—electricity. Televised documentaries on Shippingport spend considerable time explaining the nuclear process with special attention to safety and quality control. In Westinghouse's documentary, attention is given to the *simplicity* of the nuclear miracle; so, as wondrous and peaceful as atomic power is, it is presented as comprehensible and thus manages to straddle the delicate line between being mystical and fathomable.[23] At the time, Shippingport was presented as a pioneer, and perhaps it was the marketable success of this televised phenomenon in the 1950s that secured the way for other plants in the Pennsylvania area—including Three Mile Island.

for nuclear power. The mere opening of a plant would not provide much entertainment; however, a neutron rod and a Presidential sorcerer would provide a much-needed spectacle. As rhetoric goes, the moment was strong and merely cemented the message that had been fed to audiences throughout the twentieth century—technology in the hands of Americans is miraculous . . . but now also benevolent.

Figure 2.1 President Eisenhower and the 'neutron rod' (1954). Eisenhower uses a 'neutron rod' to start construction on the Shippingport Atomic Power Station

'Untitled: HD.3C.029', in *Energy.Gov* (Flickr page of United States Department of Energy) (2013)

The new benevolence of the atom was also explored in fictional films of the 1950s. One key example is *Atomic Energy as a Force for Good* by The Christophers; this film attempts to present nuclear power as lifesaving and a gift from God.[24] Starring Paul Kelly as John Vernon, the fictional but educational drama with dominant Christian themes depicts Vernon traveling with his ill grandchild; the child has a brain tumor and is on her way to spend her final days at her grandfather's ranch. During the trip, the family witnesses an atomic bomb test and shortly after arriving at the ranch they discover a nuclear plant is to be built nearby. Initially, Vernon is horrified and compares the plant to having a bomb on his land. The town debates the coming of the nuclear plant, considering the risks and going as far as mentioning Hiroshima. The town decides to challenge the Atomic Energy Commission (AEC) to prevent the construction of the plant. However, as an AEC employee explains, the 'confusion' and 'fear' of the plant is linked to a misunderstanding of the technology due to lingering anxiety over the bomb: "In spite of Hiroshima, if we can control it rightly, as fire has to be controlled rightly, it will be one of the greatest blessings we have ever received."[25] The words of the AEC do little to reassure Vernon, who remains opposed to the plant. However, he is finally convinced of the benefits of the technology when he realizes that atomic research in the field of medicine can help save his granddaughter. The film concludes that God has trusted us with a great power that we must now use for peace and betterment; as we are reminded that as God created the atom, it cannot truly be evil.[26]

Nevertheless, despite the publicity of the peaceful atom and the advertisement of its benefits, commercial companies did not demonstrate the enthusiasm and the large take-up initially expected.[27] William Beaver attributes this to nuclear power being "a largely unproven technology".[28] If the public was uncertain, so too were commercial companies. Concern was not necessarily over danger here, but more likely linked to the easy and traditional reliance on established energy technology (such as fossil fuel). Landmark plants such as Calder Hall (United Kingdom, 1956) and Shippingport (United States, 1957) sparked new interest in nuclear technology; however, the industry would experience

TURNKEY

Turnkey contracts offered by General Electric and Westinghouse were available in the 1960s and promised completed plants for a fixed price: all the buyer needed to do was 'turn the key'. Questions were raised about the low price of these plants, and soon it became apparent that turnkey contracts put General Electric and Westinghouse at a significant financial loss.

peaks and troughs as new reactors and experimental designs propelled the industry forward, while frequent regulatory changes and varying commercial interest slowed developmental progress.[29] While many plants were constructed, the 1960s and 1970s saw the industry still struggling to emerge with experimental designs, mounting costs and 'turnkey' cancellations problematizing the dream of the 1950s.

While nuclear power was struggling to emerge in the ways promised during the exuberance of the fifties and the magical excitement of Atoms for Peace, the 1960s was very much positioned as a 'futuristic' moment due to President John F. Kennedy's 'New Frontier' and the wonder of the Space Race. The 1960s was ripe for technological success, and there was certainly need to sell the public an exciting vision of the future. Nuclear power, of course, was very much intended to be a 'technology of tomorrow'. However, the swift rate of technological evolution for some was a shock and the dream of the nuclear remained problematized.

TECHNOLOGICAL EVOLUTION AND A MECHANIZED FUTURE IN KENNEDY'S 'NEW FRONTIER'

As I noted at the start, this chapter will engage with what can be called the 'long 1960s'. Some scholars, such as Jon Agar, a historian of science and technology, speak of "a period of change" in the 1960s as starting in the mid-1950s and stretching to the mid-1970s.[30] For Agar, the long 1960s "draws attention to some continuity of aspirations and attitudes, actions and institutions that together were seen to be part of a process of change".[31] Nuclear debate is one area that can be said to participate in the long 1960s as during this period, 1957–1974, the long 1960s saw a shift, as Andrew Blowers and David Pepper note, between "euphoria" marked by a "flood of orders" for plants before a decline of "uncertainty" in the late 1960s and early 1970s .[32] This period also saw increased public scrutiny and saw the development of anti-nuclear protest shifting towards energy as well as weaponry.

The long 1960s was a period marked by technological wonder. The era saw a time of rich, diverse, and astonishing scientific victories: rapid computing advancements, extraordinary discoveries (James Watson, Francis Crick, and Maurice Wilkins were awarded the Nobel Prize for mapping DNA structure), and the 'Space Race' (starting with Major Yuri Alexeyevich Gagarin's first trip into space and culminating in Neil Armstrong's moonwalk in 1969). However, while America was central to many positive movements, the period also saw problematic and difficult moments. To give two examples, the glory of the Space Race was problematized by numerous fatal accidents, and America's controversial involvement in the Vietnam War caused many to question the actions of the government. The world seemed to be both full of opportunity and fraught with danger. Nuclear technology, already demonized through its link to death and contamination, faced a struggle to be conceptualized as a positive and domesticated technology during a time when everything seemed risky and uncertain. The Space Race perhaps showed the American people that risk paid off, that scientists had a strong grasp on new and radical science, that technology heralded extraordinary breakthroughs; but, at the same time, accidents and malfunctions highlighted the inherent hazards that come with 'new' technologies. Do the benefits of technological progression outweigh the costs? This question was at the forefront of discussion when the Windscale disaster occurred, and then repeated during the numerous nuclear accidents that marred, and even helped define, the nuclear age.

In addition, the 1960s was a complicated time of subtle yet emerging anxieties in society, and this is evidenced through the speeches of President J. F. Kennedy. In his 1960 election campaign, Kennedy spoke of the technological climate of the United States with a degree of concern. For Kennedy, the 1960s promised "unknown opportunities and perils—a frontier of unfulfilled hopes and threats".[33] Kennedy spoke of the New Frontier as beyond control: "the New Frontier is here, whether we seek it or not".[34] Kennedy's rhetoric echoes concern over the place of humanity during a time of constant technological change. Kennedy concluded his speech by referencing Isaiah 40:31 from the Christian Bible: "They that wait upon the Lord shall renew their strength; they shall mount up with wings as eagles; they shall run and not be weary."[35] Kennedy's words call for God to provide strength and comfort in an era of human displacement. The book of Isaiah involves themes of judgment and salvation. Isaiah's warning can be interpreted as a warning against artificial idols and the threat of ruin for those who place their faith in artificiality rather than in God (8.17–22). Here, we return to the apocalyptic theme of Chapter 1 as Kennedy's biblical quotation concerns the sinful nature of Babylon. Kennedy's application suggests that the situation in the United States can be perceived as dangerous; consequently, the people must trust in God. Later, Kennedy claimed: "a supreme national effort will be needed in the years ahead to move this country safely through the 1960s".[36]

Although Kennedy was not technologically pessimistic (he spoke positively about the pioneers of the future especially in regards to space exploration), he could not avoid recognizing the potential disaster the New Frontier could bring. In 1963, Kennedy described the 'New Frontier'

CUBAN MISSILE CRISIS

It was with the Cuban Missile Crisis that fears over a nuclear Cold War becoming 'hot' reached a pinnacle, as for thirteen days the Soviet Union and the United States of America were engaged in a nuclear 'stand-off'. This crisis brought into sharp focus just how close to nuclear devastation the world could get.

as "filled with both crisis and opportunity".[37] This dual attitude towards technology further articulates technological ambivalence as documented in Chapter 1 regarding nuclear technology. Kennedy's concern was compounded by Cold War and nuclear fears (especially the Cuban Missile Crisis of 1962), and even after the treaty banning nuclear tests (in space, in the atmosphere, and underwater), he remarked that "The world has not escaped from the darkness."[38]

During the 1960s, technology was a major product of society and certainly viewed as a thing of the future. The World's Fair of 1964–1965 was technologically suggestive and featured many technological marvels such as an animatronic Abraham Lincoln. Bill Young and Bill Cotter claim that the fair "was all about the promise that science, technology and free society were the keys to building a better tomorrow".[39] According to Robert W. Rydell, the fair existed to "sell the future", and thus the future was painted as mechanical.[40] As the future was *painted* mechanical, this seemed inevitable. Technological evolution is something American scientist and economist W. Brian Arthur comments on when he argues that technologies "arose as combinations of other technologies".[41] Using Darwinism, Arthur explains the development of technology as a "process by which all objects of some class are related by ties of common descent from the collection of earlier objects".[42] Arthur formulates a theory of evolution in which all technologies are related to an ancestor. This idea of technological evolution can help us understand the interconnected relationship between scientific advances occurring after 1945. Moreover, it can help us think about the ways in which scholars were contemplating the 'uncontrolled' nature of scientific progress, which was something philosopher Hannah Arendt comments on when she claims that the "world of machines" has developed to such an extent that it seems to progress beyond conscious human will.[43]

A simple illustration of technological evolution is evidenced by the Space Race. Armstrong's famous words "One small step for man, one giant leap for mankind" seemed to encompass the very ethos surrounding technological development at this time. By placing the first man on the Moon, America had ventured forth into a new age for mankind and had successfully breached the boundaries of traditional human ability. Although the 1960s were most tightly linked to what was called 'The Space Age', the technological backbone of the Apollo missions had its roots in World War II. The Aggregat A-4 (known as the V-2, and 'Retribution Weapon 2'), used during World War II to catastrophic effect, was revolutionary as it used rocket technology and acted as a pilotless aircraft. The V-2 became notorious for causing thousands of deaths in the last months of the war alone. The history behind the V-2 was also shocking. The Nordhausen rocket factory (known as Mittelbau-Dora/Camp Dora), responsible for the development of the V-2, was in operation between 1943 and 1945. When the Americans located Camp Dora, they were horrified to discover malnourished and tortured prisoner workers and mounds of dead bodies. The atrocities at Dora contributed to a general sense of advanced technology as apocalyptic. As Michael J. Neufeld argues: "Mittelbau-Dora embodied one of the twentieth century's most horrifying lessons: that advanced industrial technology is perfectly compatible with barbarism, slavery, and mass murder."[44] Yet, despite the horrific history associated with the V-2, it fed into many peacetime developments, such as the space rocket. After the war, the United States and the Soviet Union battled to secure the technology first. For many reasons, the Space Race arguably started with the race for V-2 technology.

Apollo 11 (especially) revitalized the rocket in the same way nuclear energy was going to attempt to reposition nuclear technology as a friendly genie rather than a demonic fiend. So, as we enter the timeline of the construction and development of Three Mile Island, which started in 1967, we arrive at a time of technological optimism and scientific excellence. If America could put a man in space in 1961 and could put a man on the Moon in 1969, they could do anything. If the V-2 could evolve and be reborn as the Apollo 11, then nuclear technology could be recycled into benevolent energy. While nuclear unrest surrounding bomb testing continued to bubble away, the Space Race and other events diverted attention and seemed to prove that a time of technological mastery was upon America. Nuclear power stations were one example among many of how fearsome technologies could be rehabilitated.

However, due to the rapid rate of technological evolution and the widely different reactions to these advances, some major innovations, such as nuclear technology, became linked to social shock. Futurist Alvin Toffler's bestseller *Future Shock*, published in 1970, addresses the potential stress caused to the psyche by rapid change. In order to contend with living in a time of shock, in which evolving technological developments can be jarring (in both a positive and negative sense), Toffler urged that preventive measures be taken by thinking carefully about the ramifications of technology before implementing it: "it is undeniably true that we frequently apply new technology stupidly and selfishly".[45] Speaking in 1970, when Three Mile Island was under construction, Toffler noted the problems with nuclear power that had yet to be adequately addressed even as the industry continued to rapidly develop: "we do not even begin to know what to do with our radioactive wastes—whether to pump them into the earth, shoot them into outer space or pour them into the oceans".[46] Toffler, towards the end of his text, speaks to the difficult oscillation in society between an almost Luddite refusal of advancement and blind determination to push forward.[47] The complication of this sentiment when examining nuclear power technology was that, depending on whom you asked, nuclear power was both an example of responsible energy sourcing but also an example of irresponsible risk. A reigning sentiment, as echoed by Toffler, is that technology (especially nuclear technology) is difficult to wield perfectly: "The horrifying truth is that, so far as much technology is concerned, no one is in charge."[48] This perspective has been substantiated somewhat with every nuclear accident since Windscale—whether this is due to hardware or human error. This further plays into what thinkers like Arendt seem to suggest about technological evolution out of control.

While society demonstrated different reactions to technological development, there was also a sense of shock due to lingering social trauma and unrest over certain breakthroughs that seemed particularly perilous—usually technologies that were difficult to understand or seemed to pose a great risk to human safety. At the 'Resistance to New Technology—Past and Present' conference in 1993, scholars outlined three fields as provoking technological resistance since 1945: computers, biotechnology, and, of course, nuclear advancements.[49] Computer scientist Herman Goldstine notes that certain technological changes, like the nuclear, will not be wholly embraced by society. Why would this be the case when the Space Race was broadly accepted? Because the Apollo missions happened at a distance (viewable on television and in newspaper) and placed the risk with a small group of individuals who were actively engaged in a competition designed to ease tensions and provoke awe and American pride. Whereas, the development of nuclear plants (although also projected to the public as benevolent technology destined to fix an energy 'crisis') seemed to represent greater risk and engender greater concern because these developments were on American soil and in domestic areas. Further, while the Apollo was able to 'rehabilitate' the V-2, it was harder for the perils of nuclear technology to be erased when nuclear testing continued to wage between the Cold War countries.

'CLOSE TO HOME': ENVIRONMENTAL SHOCK AND PENNSYLVANIA DURING THE LONG 1960S

For the people of Pennsylvania at the time, there were several crises that gave specific examples of peril 'close to home' which caused 'shock' over the apparent conflict between nature and industry that would, years later, be articulated in similar ways through Three Mile Island. One notable event was the mine disaster in Centralia, Pennsylvania. Just over sixty miles from Harrisburg (the capital city of Pennsylvania), the small town of Centralia is today a smoldering ruin. The town is infamous for the cracked pavements, sunken buildings and scenes of ruination, fire, and dereliction; but, Centralia was once a lively town with a population of just over one thousand. When a man-made fire in 1962 accidentally consumed the mine that burrowed below the community, the whole area was reduced to a perpetually smoking ghost town. Once believing they were safe in a traditional mining town, the residents soon had to worry about fire, carbon monoxide, and the destruction of architecture. It wasn't only nuclear technology that made people question their safety; as Renée Jacob notes, the incident led people to question "Were they safe in their own homes and their own backyards?".[50] Environmental concerns were raised not only over the damage to the environment by this catastrophe but also articulated through the idea that nature (the natural resource of coal) had "turned against the people".[51] This is an interesting perspective considering the human fault surrounding the crisis.

Pitting human against nature was becoming a dominant theme during the environmental debates of the 1960s, which aligned the 'reckless' human more with the destructive force of technology. In addition to this, there was also unrest regarding how the government was dealing with the disaster, with worrying reports surfacing in the 1970s and 1980s suggesting that, despite earlier attempts and claims, it was predicted that there would be little hope of extinguishing the fire. As Jacobs comments: "Until the late 1970s, there was never the sense that the government might fail or that the fire was burning out of control."[52] Conspiracy theories and conflicting opinions developed—with some suggesting the fire was exaggerated (or faked) so the government could claim the coal; others pointed the finger at waste being dumped recklessly into the pits as the cause of the problem. The situation became so perilous that in the 1980s many who had decided to stay in the 1960s finally decided to flee Centralia, leaving behind an almost derelict town with houses being demolished at a rapid rate. In his closing comments, Jacobs pinpoints the main cause of unrest for those who lived in Centralia and those who fled it: "for many the greatest agony in Centralia was the fear of the unknown".[53] The Centralia group 'Concerned Citizens' was founded in the same year the Three Mile Island crisis occurred, just months after the partial meltdown. The Concerned Citizens group, who sought to have the fire put out, met with resistance and disinterest; group member Joan Girolami claimed, "Not facing the problem wasn't going to make it go away."[54] As we shall see in the next chapter, the events of Three Mile Island in 1979 would come to reflect, in part, some of the issues faced at Centralia, especially surrounding mistrust of the authorities and the distress over radiation as another example of 'fear of the unknown'.

Centralia was one of many examples of environmental catastrophe caused by (or rather linked to) industry. Despite the 1960s typically being defined by the Space Race with attention diverted to the Moon and the wider universe, much attention was given during this time to the home planet. Jeff Sanders sheds light on why this was the case: the image of the beautiful orb of Earth as shown from the vantage point of the Moon "reminded Americans of what was at stake in the imminent threats of massive chemical pollution and nuclear apocalypse".[55] The views from the Apollo missions only strengthened affection and concern for planet Earth. In 1968, astronauts from the Apollo 8 Moon mission shared for the first time astonishing images of Earth from lunar orbit. Jim Lovell, Command Module Pilot, exclaimed "The vast loneliness is awe-inspiring and it makes you realize just what you have back there on Earth."[56] In 1969, as moving as live footage of the first moonwalk was, those back on Earth also witnessed the eerie sight of man walking over a barren and lifeless land. There was an unmistakable contrast between the swirling blues, greens, and whites of the home planet and the gray deadness of the Moon. The moonwalk occurred during a time, in the late 1960s, of nuclear accidents, oil spills, fires (such as a fire on the chemically polluted Cuyahoga river in 1969), televised reports of napalm in Vietnam, as well as floods (such as the 1963 Vajont dam catastrophe), hurricanes, earthquakes (such as the 1964 Alaskan earthquake), and extreme blizzards (1967, Chicago). The American public were starting to rally behind environmentalism more rigorously in an effort to prevent the blue marble becoming another wasteland. After Earth Day (1970), many films addressed concerns over environmental disasters. *Silent Running* (1972), for example, addresses such imbalance and conflict between technology and the environment when nuclear weapons are used to destroy bio-domes that are going to help reforest the barren planet Earth.

There is a tendency to bisect America into two periods: pre and post industry—the before being a more idealized time of unblemished land and simpler living, with post-industrialization as a time of noise, dirt, and danger.[57] Leo Marx, in *The Machine in the Garden*, explains that many critical thinkers of the nineteenth century called for a balance between nature and development: technology should be used to complement natural resources and used for "the pursuit of rural happiness while devoting itself to productivity, wealth, and power".[58] This was not only true of the nineteenth century. In the twentieth century, the desire for balance continued to be articulated; in fact, much anti-nuclear sentiment, as we shall see, is to do with a perceived imbalance between nature and technology, and the conflict and risk posed to the natural world. Nuclear technology is often depicted as disrupting the status quo with nature in a way factories and mills supposedly did not. In his afterword to *The Machine in the Garden*, Leo Marx speaks of Hiroshima: "no other event in my lifetime so effectively dramatized the nexus between science-based technological progress and the cumulative, long-term degradation of the environment".[59] Events like Three Mile Island (1979), Chernobyl (1986), and Fukushima (2011) would continue to illustrate a sense of perilous imbalance in which the scales decidedly tip against the environment.

In many discussions surrounding the binary of nature and technology, a trend can be identified in which the intrusive nature of technology is described as participating in the 'rape' of the planet. For example, in *The Rape of the Earth: A World Survey of Soil Erosion* (1939) G. V. Jacks and R. O. White look at how extensive cultivation, lack of conservation and overconsumption of resources had not only resulted in the damage to the planet but to a "severance of mankind from the soil".[60] This metaphorical 'rape' is attributed to machines: "As the countryside became depleted, machines in the towns intensified the demands made on the land; machines on the railways, roads and seas enabled the increased demands to be satisfied without difficulty."[61] Rachael Carson in *Silent Spring* (1962) links pervasive technological advancements (such as pesticides) to nature contamination. The concept of science 'raping' nature is also addressed in *Rape of the Wild* (1988), which looks at how science through technology and experimentation has victimized and violated nature. On one hand, nuclear power was positioned as a 'clean' and welcome antidote to the relentless plundering of fossil fuels. However, the stigma of the technological label was hard to shake and so nuclear technology was also perceived as defiling and blighting the land, taking Americans away from their true roots, for which they were so nostalgic in the long 1960s. It was during this period that the Three Mile Island plant was planned, constructed, operated, and experienced the partial meltdown. Cleanup, as we will see, lasted until the 1990s. Not only did Three Mile Island's development coincide with an increased focus on environmentalism, it was also constructed in an area of environmental and historical note—the Susquehanna River. The river provides the plant with vital cooling water and is distinctive to the character of area.

The Susquehanna River, on which the Three Mile Island plant is situated, is the largest river in Pennsylvania and has been carving out its route for two hundred million years.[62] The River feeds numerous streams which travel deep into the surrounding counties; it is a site of attraction, fascinating history, and adjoins cities of impressive heritage, such as Harrisburg. The Susquehanna River bisects stunning countryside and is a site ripe for hiking, camping, fishing, and sailing. The area is filled with rich and diverse wildlife including deer, ducks, eagles, owls, foxes, herons, and river otters, making it an ideal spot for preservation and visitation by wildlife enthusiasts and schools. The beauty and community importance of the site prompted a full article on the river in *National Geographic* (March 1985). Historically, the river has played a vital role in the establishment of modern communities. The river made the area a rich and prosperous place to settle.

In the eighteenth century, John Harris arrived in the area and set up as a trader near the Susquehanna River—his son, named after him, would become the founder of Harrisburg.[63] Over the centuries, the river has seen much development—from early ferry crossings to a beautification movement in the early twentieth century after the picturesque riverfront became marred by "coal flats, boat docks, and unsightly fill".[64] In Gerald G. Eggert's book on industrialization in Harrisburg, he notes how in 1846, just a couple of years before Harrisburg established its first industrial plant, Edwin Whitefield drew a landscape depiction of Harrisburg: "the lithograph portrayed a peaceful, semi-rural, idyllic residential community situated on a narrow flat along the river's edge".[65] Indeed, this depiction features in the foreground two tall trees overlooking the untouched sprawling river, dotted by one single tiny boat and scattered with vegetation filled islands overlooking the banks of houses that made up the town.

This idyllic drawing barely corresponds to the view today. Today, the presence of the nuclear plant is described by some as an unsightly addition that spoils the landscape: "Only the white, hourglass-shaped cooling towers of the Three Mile Island nuclear power plant interrupt the picturesque river scene."[66] The noise of the plant is also seen as marring the idyll, as an article in the *New York Times* noted when describing the "raucous sound" of steam being ventilated.[67] Reflecting on the impact of the Three Mile Island accident in the local area, Lonna M. Malmsheimer spoke of the iconic sight of the chimneystacks and their impact on the environment in American imagination, noting that "the abstracted forms of those particular towers are nearly as recognizable as the first icon of the nuclear age, the mushroom cloud."[68]

Here, the area previously known for its picturesque landscape is now most readily associated with a nuclear accident so iconic that it is comparable to the icon of the mushroom cloud that lingers over the industry itself. The nuclear "machine in the garden" (to coin Marx's phrase) is a phrase close to how writers at the time were describing the 'rape of the wild': an invasion of an idyll. Yet, it is important to note that the plant was not the only construction linked to the destruction of nature in the area, as Susan Q. Stranahan makes reference to the "scars" left behind through logging and the "monstrous, rusty hulk of the Bethlehem Steel Corporation's four-mile long Steelton plant".[69] Yet, as Malmsheimer notes, there is an iconic significance to the looming nuclear plant that speaks of more than industrialization and articulates something far more complex about how an area can be redefined through the imposition not just of a structure but through the dawn of an age—the nuclear age—that carries with it a new understanding of industrial risk and area dominance. With the planned construction of the Three Mile Island nuclear plant, there were concerns over greater damage to the area.

While an ideal spot for the construction of a plant, the river was also susceptible to its own problems such as heavy rainfall which increased the

Figure 2.2 The Susquehanna River. The river is framed by vibrant and flourishing vegetation. In the background, the active and inactive units of Three Mile Island look over the river

Photograph by Grace Halden

risk of flooding, and the icing of the river caused by extreme temperatures. The Susquehanna is susceptible to major floods several times a century; these are often localized flash floods with little warning. Stranahan has written perhaps the most comprehensive account of the Susquehanna River, detailing not only the fascinating history of this area and the geology and economic development surrounding the river but also the risks associated with floods and pollution. In 1972, the flooding of the Susquehanna was so severe that it overran a dike over thirty-seven feet deep and drowned the surrounding valleys.[70] The flooding was due to the tropical storm Agnes that raged in June of that year, affecting several states. As a result of storm damage and flooding, houses were destroyed, businesses were ruined, people were relocated, and infrastructure (including bridges) was damaged and farms were destroyed. Water ran like rapids through the streets and whole towns were partially submerged under mud-stained water. Some citizens who had boats for use on the river were seen rowing through the streets that were once populated by cars. When the storm hit Pennsylvania, the vast amounts of rain caused the Susquehanna to rise and flood. Cities and towns all along the river were affected. In Wyoming Valley the area was so flooded that the land in the cemeteries gave way, releasing caskets into the muddy water; this gruesome sight was witnessed as coffins washed along the river. The sustained losses here were over a billion dollars.[71] In Harrisburg, homes along the riverfront were flooded as well as downtown. Similar to depictions from Centralia of nature turning on the people, a witness to the flood declared that the "city sat silently after its defeat by the river".[72] Before we even arrive at the 1979 nuclear crisis, the 1970s had already brought catastrophe to the locals surrounding the Susquehanna. What we learn from this event was that it was not only technological disasters that posed a risk, nor was it only industry that carried emotive labels. Stranahan describes the flood of 1972 as the Susquehanna "rampage", which evocatively relates both the horror and power of natural disasters. Why then, we will come to ask, was the Three Mile Island crisis so notorious? Why has it received greater coverage than natural crises like flooding?

Stranahan actually—inadvertently perhaps—sheds light on this issue. The Susquehanna was prone to flooding; early settlers documented floods and when towns and cities were established along the banks of the river flood prevention was an important issue. Those around the Susquehanna had experience of floods and were not only prepared for a disaster but even anticipated it when weather reports warned of storms. The difference with the Three Mile Island nuclear plant was that, first, the people were told that that plant did not pose any danger and, second, the threat posed by the crisis was initially impossible for the locals to detect: "Of all the disasters to befall the Susquehanna, the most frightening was the one no one could see."[73] In the next chapter, we will look in more detail at this phenomenon of invisible threat. For now, we can start to see how even though the area surrounding the Susquehanna had a history of disaster, defense, and rehabilitation, the 1979 incident was very different to anything encountered before. We will also see, in Chapter 3, how local experience with threat, and the public's evolved capacity to overcome hazard prepared them, in part, for talks of evacuation and peril. It is interesting to note how, despite the danger posed by nature, the coming of the nuclear plant was paradoxically seen to ruin the environment while simultaneously contributing to the threat it posed. Stranahan describes the construction of the plant along the river as one of many innovations not adequately thought through which only added to the existing potential risk of the river: "as with other human activities on the river, they paid no heed to the consequences".[74]

PEACEFUL ATOM IN CRISIS: ACCIDENT, SCANDAL, AND REJECTION BEFORE THREE MILE ISLAND

In the next chapter, we will look in detail at the consequences of the Three Mile Island disaster and how it affected popular culture into the 1980s and beyond. However, Three Mile Island was not the first major nuclear accident. Like many countries, following World War II, the government of the United Kingdom started work on their own nuclear weapons and nuclear energy program. In 1947, a new atomic energy site started development. The plutonium-producing plant in Cumberland (now Cumbria) was designed for the production of nuclear weapons and by March of 1952, the first plutonium was produced at the Windscale site, known simply as 'Windscale'. A year later in 1953, construction began nearby on a commercial power station that would be known as Calder Hall (of Windscale and Calder Works), and in 1956 the site became the first full-scale, commercial nuclear power station to supply electricity to the national grid.[75] Reactor 1 was opened by Queen Elizabeth II on October 17, 1956, and promised new jobs, new industry, and a bright future. However, one year after Calder Hall's opening, a fire erupted at Windscale on October 10, 1957 after a uranium fuel element ruptured. This crisis cast into doubt the United Kingdom's nuclear power industry.

The accident, attributed to a dangerous mix of employee inexperience and faulty design, released a contaminated cloud of radioactive iodine-131 which then drifted over England and Europe. With the release of over

20,000 curies, the Windscale fire of 1957 remains the worst nuclear disaster in the history of the United Kingdom. From 1957 to the end of the 1980s, investigations, reports, and statements were made regarding the incident. While official and detailed reports would not be declassified until long after the Three Mile Island crisis, the media reported milk contamination, radioactive releases into the Irish Sea, and a rise in miscarriages, leukemia, and Down's syndrome. Revelations highlighted in the media during the 1980s suggested that the then British Prime Minister, Harold Macmillan, suppressed details of the accident in an effort to retain nuclear links to America. The whole site containing both Calder Hall and the Windscale reactors was rebranded as 'Sellafield' in 1981 (as it is now commonly known), but memories of the Windscale accident were not as easy to erase.

Around the same time as the Windscale incident in the United Kingdom, there was another nuclear disaster. The precise date of the Urals disaster in the Soviet Union is unknown but many date the accident to 1957 or 1958.[76] Although Zhores A. Medvedev comments that initially there was some confusion surrounding whether a reactor had exploded (CIA claim) or if an atomic bomb had (concerns of the local people),[77] the disaster apparently involved an explosion at a nuclear waste site. However, it would take several more years for details (however vague) to become known to the wider world and public. In fact, it was in the same year as the Three Mile Island incident that an English translation of Medvedev's *Nuclear Disaster in the Urals* was published, containing numerous (previously classified) documents on the disaster. This text, based on the author's previous article in a 1976 publication of *New Scientist*, responded to criticism that the Urals event had not happened.[78] Medvedev's text sheds light on claims made by Soviet scientists in the late 1950s that a 'secret' nuclear disaster had occurred in the Urals, which had caused the deaths of hundreds and produced extensive ecological damage and environmental contamination. Medvedev not only comments on the secrecy of the Soviet government surrounding the silencing of the crisis but the secrecy of the American CIA as well (presumably, he suggests, to limit public concern over nuclear power).[79] However, in the 1970s word was getting out. Not only was the *New Scientist* publishing on the event but mainstream newspapers such as *The Washington Post* responded to declassified papers on the issue. This information was the CIA's report, released to campaigner and political activist Ralph Nader under the Freedom of Information Act.[80]

Windscale and the Urals were at a distance though; they were nuclear incidents of note, but they did not directly impact America. However, in 1961 America experienced its first significant and publicized nuclear accident, just five years after Windscale. In January 1961, in Idaho, three servicemen were killed instantly when a fault was encountered with control rods at the Stationary Low-Power Reactor (SL-1). Radiation was so lethal during this incident that the deceased were buried in coffins lined with lead and the investigation into the accident had to be postponed for months to allow the radiation levels to diminish.[81] Due to the fatalities at the plant, the AEC presented an educational documentary 'The SL-1 Accident'. The 1960s and 1970s saw problems at other plants in America including, but not limited to, the famous Browns Ferry Nuclear Power Plant incident in Alabama. This crisis involved the accidental starting of a fire which spread and lasted for hours after a worker searched for air leaks with a candle. In 1976 John Fuller published *We Almost Lost Detroit*, a title inspired by an Enrico Fermi worker's exclamation "let's face it, we almost lost Detroit" in response to the 1966 accident in which the fuel core of the Enrico Fermi experimental breeder reactor partially melted.[82] *We Almost Lost Detroit* was a Reader's Digest book published for the public and its front cover ominously warned, "This is not a novel. This is the shocking, terrifying truth! Could Hiroshima happen here? It could. It almost did."[83]

A link between domestic, peaceful uses of atomic technology and imagery of explosive weaponry was seen at the same time during the Plowshare program. Plowshare (also known as Operation Plowshare) developed from the Atoms for Peace project as an example of how the government and military could employ nuclear weapons for peaceful purposes to benefit civilian populations. The project was developed by the AEC, the University of California's Radiation Laboratory, and the Lawrence Livermore National Laboratory. This project ran during the long 1960s and concentrated on presenting nuclear technology, namely nuclear explosives, in a positive, safe, and beneficial way. In American publication *Science*, it was claimed that the motive behind Plowshare was to "free the atom from its deep associations with Hiroshima" as well as to promote the technology as potentially beneficial and peaceful.[84] Between 1957 and 1973, the long 1960s, thirty-five explosions occurred as part of the operation.[85]

In a production by the AEC, the Plowshare program is explained by narrator John Caple as the peaceful use and harnessing of the nuclear bomb:

> To bring water and food where there is only parched earth; and people where there is desolation. To bring freedom of movement where there are imposing barriers; and commerce where nature has decreed there will be isolation. To bring forth a wealth of materials where there are vast untapped resources; and a wealth

> of knowledge where there is uncertainty. To perform a multitude of peaceful tasks for the betterment of mankind, man is exploring a source of enormous potentially useful energy: the nuclear explosion.[86]

The ambition of Plowshare was to explode land ('earth breaking' or 'earth moving') in order to utilize the earth in new ways. For example, the crater produced by a bomb blast could act as a reservoir to hold rainfall. The AEC explained that Plowshare could help recover natural resources as nuclear explosions could expose resources humans have been unable to reach including gas, oil, metals, and other important materials. Strategic craters could help create routes through difficult terrain such as mountains. Plowshare also wanted to use nuclear excavation to help construct canals, railroads, and dams. Nuclear weapons were presented as a quicker, cheaper, and more efficient method than the many tons of TNT that would be needed to achieve a similar result. While the project was presented as an ideal route through which to both forward science and positively utilize the earth, Plowshare was also an exercise in trying to steer public opinion towards embracing nuclear experimentation and testing during a time of skepticism.[87] Plowshare was presented as the perfect combination of nuclear technology and a peace mission. Moreover, it was presented as not only beneficial for the people but completely safe.[88] By exploding bombs deep in the earth as part of their excavation work, the idea was that the safety issues surrounding blast damage and radiation would be nullified. Further, above ground explosions would be conducted in remote locations for the benefit of the experiment and were said to pose no threat to residential areas. The AEC documentary on Plowshare also boasted the unity of many different groups and factions—from geology to radiobiology and beyond—giving the impression of general scientific acceptance of this program.

The project's first venture and demonstration of cratering through nuclear explosion was on July 6, 1962, as part of Project Sedan. The explosion occurred at the Nevada test site and while the crater produced reflected scientific predictions, the dust cloud and fallout produced was far greater than anticipated.[89] The Sedan project culminated with a crater 320 foot in depth and 1,280 foot in diameter following the explosion of 104 kilotons.[90] In the June 1963 issue of *Life*, the magazine presented a story on the work being done by Plowshare featuring a full-page image of the crater in Nevada, the size of which was highlighted by the tiny scientists standing in the crater's flattened center no bigger than pinpricks. Reiterating the ambition of the project, *Life* also explained to readers how Plowshare and the AEC were attempting to "develop peaceful uses for nuclear explosions".[91] However, the larger than expected dust clouds traveled further than anticipated: "it drifted over Colorado, Wyoming, Nebraska, and the Dakotas (and probably into Canada), before turning south of Chicago on an eastward route to the Atlantic Ocean".[92] Despite claims of limited fallout, Scott Kirsch notes that scientists challenged the AEC after detecting radiation in the field beyond official parameters; one scientist to take a stand was the director of Radiological Health from the University of Utah, Robert Pendleton.[93] Once again the idea of a 'nuclear solution' morphed into yet another 'nuclear problem' leading to criticism by public and expert groups. Atoms for Peace was not, in fact, *peaceful*.

In 1963, the limited nuclear test ban treaty was signed by the United States, the Soviet Union, and Great Britain; this treaty not only limited nuclear testing but also the 'peaceful' explosions as well. In Article 1 of the treaty, the rules are clear: "Each of the Parties to this Treaty undertakes to prohibit, to prevent, and not to carry out any nuclear weapon test explosion, or any other nuclear explosion, at any place under its jurisdiction or control."[94] The treaty agreement significantly hindered the Plowshare operation but did not shut it down; in fact, the Plowshare Palanquin project in April 1965 was claimed to have not *explicitly* violated the agreement.[95] Palanquin was also significant as the AEC made a concentrated effort to suppress public knowledge of the experiment until days after the detonation.[96] The impression this event gave was that the treaty, agreed only two years earlier, was being pushed to the limit and the powers that be—the AEC and the government—were intentionally withholding important information regarding nuclear experimentation from the general public. Such incidents only fueled suspicion and unrest.

The project did find support from people like Ralph Sanders who asserted that the Plowshare project enabled experimentation that would help "test ideas" in the pursuit of attempting to use nuclear explosives peacefully.[97] Nevertheless, Plowshare started to wind down in the 1970s and was terminated in 1977, the same time nuclear power expansion ground to a halt in the United States.

Before its termination, the Plowshare project made its way to the Susquehanna River. In 1967, the same year construction of Three Mile Island Unit-1 began, Plowshare brought Project Ketch to Harrisburg. Stranahan explains that the choice of location may have been down to Pennsylvania's first commercial atomic energy plant, Shippingport, which may have implied local acceptance of the project through familiarization with peaceful nuclear technology.[98] However, this was not the case. Severe opposition to the proposed atomic blasts prevented the project's commencement. Stranahan notes that over 25,000 signed a petition against the project; the major concerns seemed to involve anxiety over radiation, the safety of the local community, and worry about damage to the environment.[99] Despite the presence of nuclear power stations in Pennsylvania and the duration of Plowshare in the long 1960s, not a single bomb detonated in the State: Plowshare was not welcome. However, this is not to say Pennsylvania rejected nuclear technology at this time; indeed, as Stranahan mentions, Plowshare was brought to the State because it was believed the people would embrace it.

During the long 1960s, Pennsylvania saw the construction of many plants with more planned along the river. Stranahan notes that the 'rosy reports' issued by the authorities about nuclear power had led most to believe, without question, the claims of the experts.[100] But the rise of power stations in the State and the plans for more along the river shook the pre-existing sense of ambivalence as people started to question more openly the safety aspect of these developments—just as they did with Plowshare. Growing concerns and unrest led to hearings in 1970 by the Pennsylvania Senate at Harrisburg, where experts debated the risks and benefits of the technology as well as outlining what they believed were the dangers. Many experts, including J. L. Everett, spoke out openly about the mastery they had over radiation and reiterated that safety was something entirely ensured. In 1973, the *Delaware County Daily News* spoke of a shift in nuclear power reception, noting that nuclear technology shifted from godly to something that provoked concern over danger to humans and the environment.[101] This local newspaper reported on the 1973 release of the suppressed 1965 document of public safety and nuclear technology 'Theoretical Possibilities and Consequences of Major Accidents in Large Nuclear Power Plants.'[102] The article concludes that although opinion on the safety in plants varies from expert to expert, the biggest breakthrough is that debate is now happening publically which is better than the " 'Papa-knows-best' approach" in which decisions are secretly made by the authorities.[103]

'Papa', many started to think, did not know best; in fact, for some, 'Papa' could not be trusted at all. One of the greatest conflicts to trust was seen with the Karen

WASH-740, 'THEORETICAL POSSIBILITIES AND CONSEQUENCES OF MAJOR ACCIDENTS IN LARGE NUCLEAR POWER PLANTS'

The 1957 Atomic Energy Commission's Wash-740 report looked at the potential for nuclear power accidents and concluded that the worst feasible accident could result in 3,400 deaths.[104] Many consider this report to have been commissioned and diluted to "reassure the public" that anxieties were baseless[105]; however, reports in the 1970s quoted higher levels of fatalities but these reports were suppressed.[106]

Silkwood case. In 1974, Karen Silkwood was found dead in her car on the way to meet a reporter. Karen, a worker at Kerr-McGee and member of the Oil, Chemical, and Atomic Workers' Union, became concerned over safety at the Cimarron site in Oklahoma. Although Silkwood joined the nuclear industry in 1972 as optimistic as any worker embarking on a new career in an exciting new field, she soon became disquieted over problems at the plant. In July 1974, Silkwood discovered she had been contaminated with radiation. Silkwood believed she had been intentionally exposed to plutonium as a warning against her involvement with the union and her outspokenness regarding safety problems at Kerr-McGee. The union arranged for her to share her story and the documents she had gathered against Kerr-McGee with *The New York Times*. However, in November 1974, on the way to meet with a reporter, Silkwood died in a car crash. While the union claimed she had been intentionally run off the road, the law ruled that Silkwood had crashed her car after becoming intoxicated. In 1976, Silkwood's father and children decided to civilly sue Kerr-McGee for negligence over Silkwood's contamination; the trial commenced in 1979, the year of the Three Mile Island crisis. The family was initially awarded over ten million dollars (eventually, an out of court settlement was reached in 1986 for $1.38 million, see Chapters 3 and 4). Outside court, the Silkwoods' lawyer Gerald Spence addressed both the public and industry: "It sends a message to the government and to the nuclear industry that they have to tell the truth."[107] The trial brought to public attention the story of Silkwood, a story which, in the 1980s, would become an Academy Award nominated Hollywood blockbuster starring Kurt Russell, Cher, and Meryl Streep (*Silkwood*, 1983).

Seemingly, the promise of the 1940s and 1950s of nuclear technology becoming a peaceful 'savior' was not necessarily materializing for the public as accidents, unrest over testing, and scandals received focused media coverage. Furthermore, in the 1970s, commissioned plants were canceled, constructions were stopped part way through, and areas promised a surge in jobs with the development of a new plant were finding such dreams were not going to be realized. Long before the Three Mile Island accident, nuclear power development appeared to have hit problems; during the 1970s, approximately forty nuclear power plants had been canceled with only fifty-three out of 129 existing projects reaching completion.[108] To the public, there seemed to be instability in the nuclear market that would have been difficult to align with the extremely optimistic assurances of the government in the 1940s and 1950s. Emotive and troubling accidents like Windscale and scandals like Silkwood's suspicious death may have become, for some members of the public, a reason for this instability. In reality, William Beaver attributes the decline in the nuclear industry not to accidents but to the industry's rocky start surrounding the development and cost of the technology.[109] By 1977 nuclear power expansion had halted; plans were scrapped, some builds abandoned, and some converted to other purposes—the benevolent Nuclear Genie was returning to his bottle. Three Mile Island became operational during a time when skepticism was on the rise and the nuclear dream was nearing an end. While the crisis in 1979 came after the industry ceased new projects, the Three Mile Island accident would linger as a chief reason for the bottling of the Nuclear Genie. In many respects, the Pennsylvania crisis confirmed that the cessation of development was appropriate.

AN AMERICAN CRISIS

The announcement that Pennsylvania was to gain another nuclear power plant was announced nationwide by the press. *The Washington Post, Times Herald* in February 1967, in a tiny snippet, ran a one-paragraph story 'Susquehanna Island to Be A-Plant Site'.[110] The *New York Times* also announced 'Atom Power Plant Planned for Island in Pennsylvania' on the same date with little elaboration and buried on page thirty-six.[111] When the announcement came that a second unit had been approved at the Three Mile Island site, national coverage was again sparse; the *Wall Street Journal* mentioned the development on page ten in a succinct couple of paragraphs.[112] However, in local reports plant construction received more attention as new plants meant new jobs and new opportunities for those living near the area. Consequently, before opening, Three Mile Island was becoming popular in local advertisements. In 1967, Gilbert Associates started to advertise in local newspapers for experienced drafters promising involvement with "exciting projects" including Three Mile Island.[113] The nuclear plant was pitched as a project symbolizing the future, with Gilbert Associates encouraging applications by guaranteeing "involvement in the future" through working in a "modern" job in the nuclear power industry.[114]

While national presses did not have a lot to say about the establishment of more nuclear plants, *The Washington Post, Times Herald*, for example, covered the nuclear debate in Virginia, Chesapeake Bay, Midwest, and New England surrounding the risk posed by nuclear plants. The article also covered citizen unrest over the proliferation of plants which were seen to harm the environment and the "aesthetics" of areas with historic interest and value; special mention was given to Pennsylvania and the construction of Three Mile Island as an example of an area becoming saturated by plants.[115] Local papers also reported on setbacks, accidents, and delays. One such report in *The Record-Angus* from Greensville,

Pennsylvania in 1971 employed the ominous headline 'Island Scene of Blaze'.[116] This story involved a fire which affected trailers, damaging employees' personal belongings and having little to do with the functioning of the plant. In a damning article in 1976, the *New York Times* revealed that there had been approximately 175 threats posed to the nuclear power industry since 1969—including hoaxes and bomb threats.[117] Some of these threats involved Three Mile Island, including bomb scares (September 1972, and March and November 1973).[118] One threat, David Burnham reported, involved "a disturbed former employee" breaking into the Three Mile Island plant in 1976, provoking concerns over security at the facility.[119]

However, once the plant was open and active, reporters were also quick to comment on the successes of Three Mile Island. In 1976, the county newspaper *Wellsboro Gazette* published an article detailing how Three Mile Island had been highly ranked for its dependability, noting the plant was even classed as first in the country.[120] Two years later, one year before the Three Mile Island crisis, the same paper reported optimistically about the forthcoming operation of Unit-2, claiming that the operation would add substantial kilowatts to help the country respond to growing energy needs.[121] Despite some minor setbacks, the opening of the plant was a momentous occasion. During the opening ceremony for Three Mile Island in 1978, Energy Secretary John F. O'Leary publically announced that "nuclear power is a bright and shining option for this country". It would be in March 1979, though, that Three Mile Island would be promoted from snippet articles and minor stories as O'Leary's "bright and shining" nuclear future was met with the harsh reality of America's worse commercial nuclear disaster to date.

The accident happened in 1979 but was part of a much longer story, and many of the issues articulated during and after the accident resonated with concerns expressed much earlier. In this chapter, we have seen how the long nuclear 1960s (1957–1974) straddled major breakthroughs and concerns. The calming presence of the Atoms for Peace project was problematized by accidents like the 1957 Windscale incident and unpopular programs like Plowshare. The magic and mystery of the atom, as promoted during Atoms for Peace, did not necessarily translate, as nuclear development ground to a halt two years before the Three Mile Island crisis. Accidents and crises were better tolerated when they were at a distance (the United Kingdom and the Soviet Union) and involved risk to a small group (Space Race); but, when industry calamities (such as Centralia), nuclear accidents (such as SL-1, 1961), and experimental programs (Plowshare) posed threat to civilian communities in America, protests were more common. Toffler and Goldstine both describe the 1970s as a comparable time to the Industrial Revolution in which society may react like the Luddites and attack the technology.[122] For Dominic Sandbrook, the 1970s in America saw "hinge moments" which marked a "transition between an old world and a twenty-first century world" in terms of technological development and "reinvention".[123]

Nuclear technology, it seemed, would always be a "hinge" moment and represent a struggle for transition. While the V-2 could be rehabilitated through the Apollo missions, the nuclear plant would always be complicated, to some degree, by its past in 1945, and also by the ongoing thorny issues of radiation, accidents, scandals, and protest. In the next chapter, we will see how Three Mile Island confounded the idea of the peaceful atom; the magic was over and the earlier shadow of Windscale and the Urals fell over Pennsylvania and Americanized the nuclear threat. We will see how, even though the plant was established during a time of technological boom, there was suspicion regarding how successful human mastery was over these advances. Technical mishaps and human error would call into question the implementation of dangerous technologies that inexperienced engineers and flawed systems struggled to control. The 'New Frontier' brought with it many wonders, but these wonders were not always completely 'in hand'; the Nuclear Genie was not always completely under control. Anger over misinformation, confusion over rumors, and frustration over incompetent responses to disaster were articulated during the Three Mile Island incident, just as they were during the Centralia crisis. Three Mile Island would not be the first incident in which the people challenged the authorities and expressed frustration over slow and incorrect information. It was also not the first accident to cause concern over the environment and the local populace. And it was not the first to cause shock. It also would not be the last. But Three Mile Island would become a landmark event alongside which much greater nuclear disasters would be discussed. Windscale was the problem of the United Kingdom, Chernobyl was the catastrophe of the Soviet Union, and Fukushima was the tragedy of the Japanese—but Three Mile Island was an American problem. Three Mile Island was a crisis for the Americanized peaceful atom and, while this crisis was ultimately averted, the shock waves of this near miss are still felt today.

NOTES

1 Albert Einstein 'Letter to F. D. Roosevelt', August 2, 1939, in *AtomicArchive*, available online at www.atomicarchive.com/Docs/Begin/Einstein.shtml (accessed August 18, 2016).

2 Einstein, 'Letter to F. D. Roosevelt'.

3 Einstein, 'Letter to F. D. Roosevelt'.
4 H. D. Smyth, *A General Account of the Development of Methods of Using Atomic Energy for Military Purposes under the Auspices of the United States Government 1940–1945* (Washington D.C., 1945), p. 29.
5 'The Atomic Energy Act', United States Nuclear Regulatory Commission, August 30, 1954, p. 15.
6 'Project "Candor"', Security Information Secret, Unclassified (July 22 ,1953), in *The Dwight D. Eisenhower Library* (n.d.), available online at www.eisenhower.archives.gov/research/online_documents/atoms_for_peace/Binder17.pdf (accessed July 19, 2016).
7 *The American Atom. A Documentary History of Nuclear Policies from the Discovery of Fission to the Present 1939–1984*, ed. by Robert C. Williams and Philip L. Cantelon (Philadelphia: University of Pennsylvania Press, 1984), p. 73.
8 Stewart Alsop, 'Eisenhower Pushes Operation Candor', *The Washington Post*, September 21, 1953.
9 *The American Atom*, p. 74.
10 Harry S. Truman, 'Statement by the President of the United States', Washington D.C., The White House, August 6, 1945, in *Harry S. Truman Library & Museum* (2016), available online at www.trumanlibrary.org/whistlestop/study_collections/bomb/large/documents/index.php?documentid=59&pagenumber=1 (accessed July 19, 2016).
11 *The Atomic Age Opens*, ed. by Donald Porter Geddes and Gerald Wendt (New York: World Publishing Company, 1945), pp. 29–30.
12 Smyth, *A General Account*, p.164.
13 'The Bomb that Won the War', *Science Comics*, 1, (Springfield: Humor Publications, January 1946), p. 7.
14 Dwight D. Eisenhower, 'Atoms for Peace', 470th Plenary Meeting of the United Nations General Assembly, December 8, 1953, in *IAEA* available online at www.iaea.org/about/history/atoms-for-peace-speech (accessed July 14, 2016).
15 Eisenhower, 'Atoms for Peace'.
16 *A is for Atom*, dir. by Carl Urbano (General Electric Company, 1952).
17 *Power and Promise: The Story of Shippingport*, The United States Atomic Energy Commission (Mode-Art Pictures, c. 1950).
18 'It's Electric!', *The Atom for Peace*, c. 1950(58?).
19 'It's Electric!', *The Atom for Peace*.
20 Joyce Nelson, *The Perfect Machine* (Ontario: Between the Lines, 1987), pp. 44–45.
21 *Power and Promise: The Story of Shippingport*, The United States Atomic Energy Commission (Mode-Art Pictures, c. 1950).
22 *Power and Promise: The Story of Shippingport.*
23 See *Atomic Power at Shippingport* (Westinghouse Electric Corp, c. 1950).
24 *Atomic Energy as a Force for Good*, dir. by Robert Stevenson (The Christophers, 1955).
25 *Atomic Energy as a Force for Good*, dir. by Stevenson.
26 *Atomic Energy as a Force for Good*, dir. by Stevenson. The Christophers, from which A Christopher Production came, were a Christian inspirational group founded by Father James Keller and Keller also features in *Atomic Energy Can Be a Blessing* (1952) speaking of atomic technology as a divine offering.

27 William Beaver, 'The Failed Promise of Nuclear Power', *Independent Review* 15.3 (2011), 399–411.
28 Beaver, 'The Failed Promise of Nuclear Power', pp. 399–411.
29 For a concise, yet detailed analysis, of reactor development and the nuclear industry in the 1950s see William Beaver, 'The Failed Promise of Nuclear Power'.
30 Jon Agar, 'What happened in the sixties?', *The British Journal for the History of Science* 41.4 (2008), 567–600 (p. 567).
31 Agar, 'What happened in the sixties?, p. 569.
32 *Nuclear Power in Crisis*, ed. by Andrew Blowers and David Pepper (London: Croom, 1987), p. 102.
33 John F. Kennedy: 'Address of Senator John F. Kennedy Accepting the Democratic Party Nomination for the Presidency of the United States', Memorial Coliseum, Los Angeles, July 15, 1960, in Gerhard Peters and John T. Woolley, *The American Presidency Project* (2016), available online at www.presidency.ucsb.edu/ws/?pid= 25966 (accessed July 16, 2016).
34 Kennedy, 'Address of Senator John F. Kennedy Accepting the Democratic Party Nomination for the Presidency of the United States'.
35 Kennedy, 'Address of Senator John F. Kennedy Accepting the Democratic Party Nomination for the Presidency of the United States'.
36 John F. Kennedy, 'Acceptance Speech', Hyannis Armory, Hyannis, Massachusetts, November 9, 1960, in *John F. Kennedy: Presidential Library and Museum* (n.d.), available online at www.jfklibrary.org/Research/Research-Aids/JFK-Speeches/ Hyannis-MA-Acceptance-Speech_19601109.aspx (accessed August 12, 2012).
37 John F. Kennedy, 'Remarks at the Dedication of the Aerospace Medical Health Center', San Antonio, Texas, November 21, 1963, in *John F. Kennedy: Presidential Library and Museum* (n.d.), available online at www.jfklibrary.org/Research/ Research-Aids/JFK-Speeches/San-Antonio-TX_19631121.aspx (accessed August 12, 2012).
38 John F. Kennedy, 'Address Before the 18th General Assembly of the United Nations', United Nations, September 20, 1963, in *John F. Kennedy: Presidential Library and Museum* (n.d.), available online at www.jfklibrary.org/Research/Ready-Reference/JFK-Speeches/Address-Before-the-18th-General-Assembly-of-the-United-Nations-September-20–1963.aspx (accessed August 12, 2012).
39 Bill Young and Bill Cotter, *Images of America: The 1964–1965 New York World's Fair* (Chicago: Arcadia, 2004), p. 13.
40 Robert W. Rydell, *World of Fairs: The Century-of-Progress Expositions* (London: The University of Chicago Press, 1993), p. 116.
41 W. Brian Arthur, *The Nature of Technology: What It Is and How It Evolves* (London: Penguin, 2010), p. 2.
42 Arthur, *The Nature of Technology*, p. 15.
43 Hannah Arendt, *The Origins of Totalitarianism* (London: Harvest, 1976), p. 153.
44 Michael J. Neufeld, 'Introduction', in *Planet Dora: A Memoir of the Holocaust and the Birth of the Space Age*, by Yves Béon, trans. by Yves Béon and Richard L. Fague (Westview Press: Colorado, 1997), pp. ix-xxviii (p. ix).
45 Alvin Toffler, *Future Shock* (New York: Bantam, 1971), p. 218.
46 Toffler, *Future Shock*, p. 218.
47 Toffler, *Future Shock*, p. 219.

48 Toffler, *Future Shock*, p. 220.
49 *Resistance to New Technology: Nuclear Power, Information Technology and Biotechnology*, ed. by Martin Bauer (Cambridge: Cambridge University Press, 1997), Preface.
50 Renée Jacobs, *Slow Burn: A Photodocument of Centralia, Pennsylvania* (Pennsylvania: Pennsylvania University Press, 2010), p. xiv.
51 Jacobs, *Slow Burn*, p. xv.
52 Jacobs, *Slow Burn*, p. xvi.
53 Jacobs, *Slow Burn*, p. 149.
54 Jacobs, *Slow Burn*, p. 101.
55 Jeff Sanders, 'Environmentalism', in *The Columbia Guide to America in the 1960s*, by David Farber and Beth Bailey (New York: Columbia University Press, 2001), pp. 273–281 (p. 273).
56 'Earthrise', NASA (2015), available online at www.nasa.gov/multimedia/image gallery/image_feature_1249.html (accessed April 13, 2016)
57 There is a degree of idealization in modern and contemporary literature which reflects nostalgically on early rural America. The popularity of American shows during the long 1960s draws attention to this nostalgia as television programs like *Rawhide* (1959–1966), *Bonanza* (1959–1973), and *Gunsmoke* (1955–1975) showed a fondness and nostalgia for untouched rural America and a return to a more simplistic and traditional heritage. Popular family shows in the 1970s like *The Waltons* (1972–1981) and *Little House on the Prairie* (1974–1983) also explored the dynamics of the traditional family unit in early America, evidencing an enjoyment in returning to the frontier way of life. It was during the long 1960s that American writer John Steinbeck spoke of the freedom and beauty of the American landscape in his account of American travel. Steinbeck, Nobel Prize winner for Literature and respected for his famous American works such as *Of Mice and Men*, embarked on a quest to revisit rural America in an effort to experience true and genuine American life. Thus, there is a nostalgia for pre-industrial America in modern literature. John Steinbeck, *Travels with Charley: In Search of America* (London: Mandarin Paperbacks, 1991).
58 Leo Marx, *The Machine in the Garden: Technology and the Pastoral Ideal in America* (Oxford: Oxford University Press, 2000), p. 226.
59 Marx, *The Machine in the Garden*, p. 369.
60 G. V. Jacks and R. O. White, *The Rape of the Earth: A World Survey of Soil Erosion* (London: Faber and Faber, sixth impression 1949), p. 281.
61 Jacks and White, *The Rape of the Earth*, p. 281.
62 Ruth Hoover Seitz, *Susquehanna Heartland* (Harrisburg: RB books, 1992), p. 6.
63 Richard H. Steinmetz, Sr. and Robert D. Hoffsommer, *This Was Harrisburg* (Pennsylvania: Stackpole Books, 1976), p. 22.
64 Steinmetz and Hoffsommer, *This Was Harrisburg*, p. 162.
65 Gerald G. Eggert, *Harrisburg Industrializes: The Coming of Factories to an American Community* (Pennsylvania: Pennsylvania University Press, 1993), p. 3.
66 Susan Q. Stranahan, *Susquehanna, River of Dreams* (Maryland: John Hopkins University Press, 1993), p. 32.
67 B. Drummond Ayres, 'Three Mile Island: Notes from a Nightmare; Three Mile Island: A Chronicle of the Nation's Worst Nuclear-Power Accident', *New York Times*, April 16, 1979, p. A1.

68 Lonna M. Malmsheimer, 'Three Mile Island: Fact, Frame, and Fiction', *American Quarterly*, 38.1 (1986), 35–52 (p. 35).
69 Stranahan, *Susquehanna, River of Dreams*, p. 145. It is important to note that the Susquehanna River has also been polluted with waste from sewage and the mines, and chemicals from cropland runoff. Noted in: Peter Miller, 'Susquehanna: America's Small-Town River', *National Geographic*, 167.3, (March 1985), 352–383 (pp. 357, 383).
70 Stranahan, *Susquehanna, River of Dreams*, p. 117.
71 Stranahan, *Susquehanna, River of Dreams*, p. 134.
72 Maria Csala quoted in Jack Brubaker, *Down the Susquehanna to the Chesapeake* (Pennsylvania: The Pennsylvania State University Press, 2002), p. 60.
73 Stranahan, *Susquehanna, River of Dreams*, p. 183.
74 Stranahan, *Susquehanna, River of Dreams*, p. 184.
75 Other plants are often cited as the first civilian power plant (for example, an Experimental Breeder Reactor in America powered a population of 1000 in 1955). However, Calder Hall is often noted as the first industrial-scale commercial plant. Shippingport, Pennsylvania, in late 1957, is known as the first commercial power plant for purely peacetime purposes; Calder Hall produced both electricity and plutonium.
76 Zhores A. Medvedev, *Nuclear Disaster in the Urals*, trans. by George Saunders (London: Angus & Robertson, 1979), p. 4.
77 Medvedev, Nuclear Disaster in the Urals, p. 16.
78 Medvedev, Nuclear Disaster in the Urals, p. 5.
79 Medvedev, Nuclear Disaster in the Urals, p. 131.
80 David Burnham, 'C.I.A. Papers, Released to Nader, Tell of 2 Soviet Nuclear Accidents', *New York Times*, November 26, 1977.
81 Jim Garrison, *The Plutonium Culture: From Hiroshima to Harrisburg* (New York: Continuum, 1981), p. 145.
82 Anna Gyorgy, *No Nukes: Everyone's Guide to Nuclear Power* (Montreal: Black Rose Books, 1979), p. 118.
83 John Fuller, *We Almost Lost Detroit* (New York: Ballantine Books, 1976), front cover.
84 Elinor Langer, 'Project Plowshare: AEC Program for Peaceful Nuclear Explosives Slowed Down by Test Ban Treaty', *Science,* New Series, 143.3611 (1964), pp. 1153–1155 (p. 1153).
85 'United States Nuclear Tests July 1945 through September 1992', United States Department of Energy Nevada Operations Office (Nevada, 2000), p. xvi.
86 *Plowshare*, United States Atomic Energy Commission (San Francisco: W. A. Palmer Films, (n.d.)).
87 See Scott Kirsch, *Proving Grounds: Project Plowshare and the Unrealized Dream of Nuclear Earthmoving* (NJ: Rutgers University Press, 2005), p. 36.
88 *Plowshare*, United States Atomic Energy Commission.
89 Kirsch, *Proving Grounds*, p. 120.
90 Recorded in: 'United States Nuclear Tests July 1945 through September 1992', United States Department of Energy, inside cover.
91 'Colossal H-Bomb Hole', *LIFE*, June 21, 1963, pp. 47–51.
92 Kirsch, *Proving Grounds*, p. 126.

93 Kirsch, *Proving Grounds*, p. 126.
94 Treaty Banning Nuclear Weapon Tests in the Atmosphere, in Outer Space and Under Water, Moscow, October 15, 1963.
95 Kirsch, *Proving Grounds*, p. 153.
96 Kirsch, *Proving Grounds*, p. 153.
97 Ralph Sanders, 'Defense of Project Plowshare', *Technology and Culture*, 4.2 (1963), 252–255 (p. 255).
98 Stranahan, *Susquehanna, River of Dreams*, p. 192.
99 Stranahan, *Susquehanna, River of Dreams*, p. 193.
100 Stranahan, *Susquehanna, River of Dreams*, p. 194.
101 Reported in 'If The Public Knew . . . Nuclear Power Plants?', *Delaware County Daily Times*, September 4, 1973.
102 Reported in 'If The Public Knew . . . Nuclear Power Plants?'
103 'If The Public Knew . . . Nuclear Power Plants?'
104 Anna Gyorgy, *No Nukes: Everyone's Guide to Nuclear Power* (Montreal: Black Rose Books, 1979), p. 111.
105 Harry Henderson, *Nuclear Power: A Reference Handbook*, 2nd edn (California: ABC-CLIO, 2014), p. 329.
106 Anna Gyorgy, *No Nukes: Everyone's Guide to Nuclear Power* (Montreal: Black Rose Books, 1979), p. 112.
107 'Karen Silkwood: A Life on the Line', host, Harry Smith (*Towers Productions*, 2001)
108 James A. Mahaffey, *Nuclear Power: This History of Nuclear Power* (New York: Facts on File, 2011), p. 128.
109 Beaver, 'The Failed Promise of Nuclear Power', pp. 399–411 (p. 409).
110 'Susquehanna Island to Be A-Plant Site', *The Washington Post, Times Herald*, February 12, 1967, p. B5.
111 'Atom Power Plant Planned for Island in Pennsylvania', *New York Times*, February 12, 1967, p. 36.
112 'General Public Utilities Units Receive Approval to Build Nuclear Plant', *Wall Street Journal*, November 3, 1969, p. 10.
113 Gilbert Associates, 'Get Involved', *Delaware County Daily Times*, June 12, 1967.
114 Gilbert Associates, 'Get Involved'.
115 Hal Willard, 'Atomic Plants Increase in Chesapeake Bay Area', *The Washington Post, Times Herald*, November 6, 1969), p. F1.
116 'Island Scene of Blaze', *The Record-Angus*, January 20, 1971, p. 1.
117 David Burnham, 'U.S. Proposes to Fine Utility for Not Keeping Unstable Ex-Employee Out of Nuclear Plant', *New York Times*, March 30, 1976, p. 26.
118 Burnham, 'U.S. Proposes to Fine Utility', p. 26.
119 Burnham, 'U.S. Proposes to Fine Utility', p. 26.
120 'Penelec Plant Ranked No.1', *Wellsboro Gazette*, September 22, 1976, p. 9.
121 'Penelec Forecasts Steady Growth', *Wellsboro Gazette*, January 25, 1978, p. 4.
122 Herman H. Goldstine, *The Computer: From Pascal to von Neumann* (West Sussex: Princeton University Press, 1993), p. 347.
123 'The 1970s', *Letters from America*, BBC, c. 1970s, available online at www.bbc.co.uk/programmes/b03zj367 (accessed April 14, 2016).

CHAPTER 3

When Science and Society Collide

The Three Mile Island Accident in Human Context

In July 1979, the Annual Homesteaders Festival at Whitneyville, Pennsylvania was held roughly 150 miles from the Three Mile Island Nuclear Generating Station. The Homesteaders Festival included community events and workshops involving crafts, baking, and farming, which directed attention to the land and self-sufficiency. Commenting on the idyllic lifestyle of living off the land, homesteader Glen Hart remarked that technology in the modern world was controlling the people and taking away a semblance of mastery over the earth and their fate. In this respect, technology seems to be paralleled with nature, and nature is positioned as more idyllic and safe in comparison. Nature is presented as part of American heritage and the backbone of a wholesome lifestyle; maybe even linked to frontier life and ideas of the American Dream. The Homesteaders Festival transported visitors back to a world pre-nuclear development and represented simplicity, domesticity, and the untainted link between settler and terrain. Hart commented that the Three Mile Island crisis, months earlier, was an example of unsecured technology and said, "We shouldn't have gone that far because we didn't have control of the technology."[1] Hart's commentary, and the nature of the festival, seemed to juxtapose unruly nuclear technology with the firmer control humans have over nature through cultivation. With cultivation the machine is 'in hand'; farming demonstrates control over the land and man-made tools. While nature can pose a threat (and indeed, Pennsylvania has experienced floods among other natural crises), the Three Mile Island predicament represented a shift in which a tool—nuclear power—became 'out of hand' with potentially catastrophic ramifications. While natural disasters are a tragedy as they are often unforeseen, unpreventable, and unprovoked, man-made disasters have accountability. Hart's statement places blame with human creators and also assigns a collective responsibility.

The crisis at Three Mile Island in March and April of 1979 was a shock. As James Mahaffey notes, "By the 1970s the United States had made it through the experimental phase of nuclear energy without any show-stopping problems."[2] Three Mile Island would not only become that 'show-stopping problem' but it would also call into question the nuclear acceptance that has been part of America since the early days of atomic culture. So influential was the Pennsylvanian crisis that Stan Benjamin of the Associated Press claimed, as reported in Maine's *Lewiston Journal*, that "the nuclear age exploded into popular consciousness on March 28, 1979"[3]—not in 1945 with Hiroshima, not in 1957 with Windscale, and not through nuclear testing. Three Mile Island gained notoriety because it was the first and largest commercial accident in American history: the crisis did not occur in a distant country, and it was not a test; this was an American incident, affecting American citizens. The Nuclear Regulatory Commission's (NRC) Harold R. Denton, director of the Office of Nuclear Reactor Regulation, noted that the uniqueness of the accident was down to the fact that there had not been an accident in twenty years and so Three Mile Island shook a sense of American "complacency".[4] Denton added that the shocking and unprecedented nature of the event made it equivalent to the "first plane crash".[5] This 'shock' helped to stunt the still new industry. In a rather poetic statement in November 1979, reporter Don Graff channeled Winston Churchill when commenting that "we may now be seeing, in the momentum provided by Three Mile Island, the end of the beginning".[6] In many ways this was true. While Mahaffey correctly notes that economic factors caused the decline of nuclear power expansion pre-Three Mile Island, it was the 'show-stopping' fears of radiation and meltdown that caused the most alarm in the local community. For the people, Three Mile Island did significantly impact how nuclear power was viewed and while Three Mile Island did not cause the momentary end of nuclear power development in the United States, it was seen to put the "last nail in the coffin".[7] In this chapter, we will explore how the media initially responded to the incident and how the event impacted nuclear optimism, fear, and ambivalence. We will come to see that Three Mile Island was as much a crisis of confidence as it was a crisis of technological failure.

THE 1979 ACCIDENT AND INITIAL MEDIA RESPONSE

"Three Mile Island was worse than the mishap portrayed in Jane Fonda's movie 'The China Syndrome'", claimed Ohio's *The Bryan Times*.[8] The film by James Bridges, released days before the Three Mile Island accident on March 16, 1979, focuses on a television reporter (Jane Fonda) and camera operator (Michael Douglas) who witness a near meltdown at a nuclear power plant. Both journalists, and the plant supervisor (Jack Lemmon), investigate the accident and discover severe and dangerous safety issues at the plant. When both parties are prevented from speaking out about these issues they take drastic action to warn the public about the threat they unknowingly face. During the film, scientists refer to a pseudo-scientific principle called 'the China Syndrome' which describes how during a nuclear meltdown, the melted fuel could sink through the earth down to China. They describe "an area the size of Pennsylvania" being wiped out if this was to occur (a strange coincidence considering the location of the Three Mile Island crisis weeks later).[9]

Regardless of the implausibility of the fictional science presented in *The China Syndrome*, the film was reviewed by some critics as believable[10] and "realistic".[11] In Thomas Tanner's review of *The China Syndrome* he even notes the educational importance of the film.[12] Physicist Frank von Hippel notes that despite errors, the film left a big impression on the public.[13] Ron Von Burg concludes, "Even if the odds of a nuclear accident are one in a million, *The China Syndrome* depicts that one time."[14] Of course, after the events of Three Mile Island the film seemed to gain a sense of credibility. This is largely why, despite factual problems, the film became important for the anti-nuclear power movement as it highlighted—in sensationalist Hollywood fashion, with an impressive (and trusted) big name cast—that the industry needed to be critiqued. It is also why so many questions directed to the authorities mentioned the film, prompting official statements to distance the factual situation from the fiction: "This is not a 'China Syndrome' type situation," plant spokesman Blaine Fabian reassured the public.[15] Yet for many, reassurances came too late as the film already prompted anxiety; as Gwyneth Cravens notes, "two weeks earlier *The China Syndrome*

KEY DATES

Three Mile Island Unit-1 began construction in 1967 and went critical in 1974. Unit-2 joined Three Mile Island in 1970 and started operating in 1978. In 1979, three months after the unit started operating, the Three Mile Island accident involving Unit-2 occurred on March 28.

had come out so I was already prepared to be terrorized by this event".[16] Bridges' film retained its hold over the people. After Three Mile Island, the local theatres UA Camp Hill and East 5 reported elevated viewing figures of *The China Syndrome*.[17] Some people, of course, did not see the film; police officer Brian F. McKay commented that he did not bother to see the film during the Three Mile Island days as he had just "lived through it".[18]

The 1979 Three Mile Island Incident

While it is not the objective of this book to address the scientific aspects of the incident at Three Mile Island, a limited amount of technical detail is required in order to provide a framework for the discussion of the incident's impact on popular culture.

So far we have spoken about nuclear power but without a real understanding of what this is. Many of us will be familiar with furnaces and boilers because we might use them in our homes. Furnaces and boilers create heat through using fuel like coal, oil, gas, and biofuels. Most of us will use a boiler and/or furnace to heat our home. However, the heat created through these systems can produce high-pressure steam and this steam can be used to power a turbine (engine), which in turn creates enough energy to power a generator, and this generator can then produce electricity. When connected to a national grid, a whole region can use this power in their homes. In a thermal power plant this is exactly what happens.

In a nuclear power plant, a nuclear reactor creates heat in what is called a 'core' but without using precious and expensive fossil fuels (coal, oil, gas). The core is full of coolant (water), and in the core a process called fission occurs through the splitting of uranium atoms by neutrons. The splitting of atoms creates phenomenal amounts of radioactive energy. A chain reaction occurs in which the atoms continue to split. Disney's *Our Friend the Atom* demonstrates this process through mouse traps weighted with ping pong balls; a single neutron (ping pong ball) sets off one mousetrap which releases more balls to set off other traps, eventually leading to hundreds of traps activating and balls flying everywhere. The culmination, Disney shows us, is the atomic explosion and resulting mushroom cloud (and the freeing of that Nuclear Genie!). However, in a nuclear power plant the chain reaction is controlled and contained (it is *peaceful*; the Nuclear Genie is restrained).

How does this control work? Dr. Heinz Haber, author of Disney's book and the documentary's host, describes the process in a nuclear power plant as an explosion that is slowed down ensuring heat and radiation are limited and contained.[19] Inadvertently, Disney, by suggesting that the nuclear core is a 'slow bomb', makes the process sound more frightening than it needs to be. In order for fission to take place, neutrons (bouncing around in the water, releasing energy) must be slowed down. If the water heats up it becomes less dense, making it not as successful at slowing down (or 'moderating') neutrons. This means that if a water-cooled and moderated reactor gets hotter, its power will naturally fall and vice versa. Ultimately, this means that temperature keeps the reactor stable. If the reactor needs to be shut down, control rods can also be dropped quickly into the core to absorb neutrons and end the chain reaction. The controlled heat produced through this system can create power through the process mentioned earlier: it creates the steam to turn a turbine which powers a generator and creates electricity. The reactor is surrounded by thick, solid concrete walls to prevent radiation leakage. The Nuclear Genie is now our friend, Disney says. However, it is essential that the core temperature is consistent because an overheated core (through a loss of water for example) can create uncontrolled heat that will build to dangerous levels. If the temperature gets too high, the fuel (uranium) can melt (known as meltdown) and the build-up of pressure can cause an explosion. The worst case scenario is that containment fails and radiation is leaked into the environment. This is what happened at Chernobyl (1986) and Fukushima (2011)—containment failure. At Three Mile Island there was a *partial* meltdown in which some of the fuel (but not all) melted.

Three Mile Island consisted of two reactors known as Unit-1 and Unit-2 and was operated by Metropolitan Edison. This plant is an example of a pressurized water reactor (PWR)—a popular style of reactor that makes up most of the commercially operating nuclear plants in the United States (the other style of reactor in America is a boiling water reactor, BWR). In a PWR system, a primary coolant loop filled with pressurized water allows the heat generated during the fission process to pass to a steam generator where water in a separate secondary loop is boiled to produce the steam needed to turn the turbine.[20] As part of the safety design of the plant, many barriers prevent humans from becoming contaminated by keeping radiation contained in the system. One important barrier is the closed cooling system that prevents the fuel melting. If the cooling system fails for some reason, there is an emergency core cooling system (ECCS) to act as a critical backup. With Three Mile Island in 1979, a malfunction in Unit-2 (Unit-1 was unaffected) caused coolant around the reactor to decrease, and operator error caused the ECCS to shut down leading to substantial overheating, severe reactor damage, and a partial meltdown: the worst commercial nuclear event in American history.

Lake Barrett, formerly of the United States Department of Energy and Nuclear Regulatory Commission, remarked that the event occurred on a regular day.[21] Edward J. Walsh described it as a "routine and boring watch".[22] However, "In the dead of night" as *Time Magazine* poetically stated,[23] at approximately 4.00 a.m. on March 28, 1979 the Three Mile Island Unit-2 plant experienced what Mahaffey terms "a series of unfortunate events" after being operational for only a few months.[24] The secondary cooling circuit that is designed to transport heat from the primary system experienced problems when one of the feedwater pumps stopped working. The malfunction caused a reduction of vital cooling water which prevented the movement of heat from the primary system. Consequently, system pressure and heat increased. Automatically, as it should do, the pilot operated relief valve (PORV) opened to deal with the pressure build-up. At this stage, the plant was dealing with the issue automatically and appropriately. This should have been the end of the problem. Unfortunately, the PORV failed to close and this went undetected. This moment marked the true start of the crisis. Colin Tucker, a Technical Specifications Specialist from EDF Energy's Sizewell 'B' Power Station in the United Kingdom, explains why the PORV was so problematic:

> The relief valve had a light on it. That light indicated what the valve had been asked to do by the control system. The control system had asked the valve to close again. So as far as the operator was concerned the valves were closed because that was what the control system had asked for. And that is such a fundamental error in the way that you design a power station. Every indication that you use should be an indication not of what you have asked the plant do, but what it actually did. In hindsight it's crazy but at the time it made for more simplistic engineering.[25]

Due to the open valve, water poured from the system and, as we know, the reactor must be cooled. So, another safety system came into play. The ECCS, as designed, started to inject the reactor with water. However, operators misread the symptoms of the ongoing crisis and thought there was too much water in the system leading them to terminate the ECCS. The system continued to lose water and steam continued to build. Reflecting on the accident, *Time* was quick to point out the problem of human error, commenting that when the operators thought the crisis was averted they were, again, mistaken.[26] The steam build-up in the primary system became so considerable that when the pumps were used, they started to vibrate as they were forced to attempt a pumping of both water and steam. To lessen the vibration, operators closed off more pumps. The water dropped below the surface of the core resulting in partial core exposure. Reactor temperature rose dangerously high, and radioactivity increased to abnormal levels. After two hours, operators realized the PORV was still open, despite the fact it should have automatically shut, and finally closed it. However, enough hydrogen had escaped into the Containment Building to give a large and rapid pressure rise when it later ignited.

It was only after sixteen hours that cooling water was able to recirculate successfully. The reactor stabilized by the close of March 28. However, the damage had been done. By the time the temperature and pressure in the system started to fall, the core had already started to melt. Early on March 30, rumors and reports that radiation had been released made it clear that the crisis was not over; residents living within ten miles of the plant were advised by Governor Richard Thornburgh to remain indoors with windows closed and air conditioning off. Shortly after this, Thornburgh closed local schools and advised women and small children within five miles to evacuate (however, far more than anticipated voluntarily abandoned the area). On the same day, reports of a developing hydrogen bubble near the reactor had many questioning whether there could be an explosion that might rupture the containment; President Jimmy Carter sent expert Harold Denton, of the NRC, to Three Mile Island to try and help get the situation under control. The authors of *TMI 25 Years Later* note that on this day concerns over the hydrogen bubble and radiation releases made the international community suddenly worried about a potential "full-fledged nuclear disaster replete with loss of life, devastating health problems, and widespread radioactive contamination".[27] On March 31, newspapers reported that the hydrogen bubble could explode, and there was disagreement between Metropolitan Edison's Vice President Jack Herbein and Harold Denton as to whether the crisis was over.[28] Anticipation of general evacuation dogged this time and many opted to abandon the area anyway. It was only on April 1, the day President Carter visited the plant, that fears started to subside. On April 4, Thornburgh announced on the *Today Show* that the crisis was considered to be at an end.

While both the Rogovin Report (an NRC investigation led by Mitchell Rogovin, 1980) and the Kemeny Report (The President's Commission on the Accident at Three Mile Island, created by President Carter and led by John Kemeny) flagged up technical and functional problems at the plant, they both highlight the "people-related problems" that plagued the crisis.[29] These 'people problems' included errors with regulation, manufacturing, organizational input, operation, management, and communication. As Daniel F. Ford points out, the mistakes made were not down to "bad luck", nor were they "freak occurrences", and long before the 1979 crisis "improper" procedures and blunders had been regularly reported to the NRC and were frequent contributing factors to nuclear industry failures.[30]

We now have a rudimentary insight into what occurred technically during the first few days of the crisis. However, of specific importance to us is the media frenzy that accompanied the crisis as approximately 400 journalists arrived at Three Mile Island hoping to cover the story. Newspapers, television, and radio became critical media through which the public was alerted to the unfolding incident.[31] While an emergency on-site was declared two hours and fifty-five minutes after the event started on Wednesday March 28, it was much later in the day when the State, NRC, and President's office were notified. It was at 8.25 a.m. that WKBO, the local Harrisburg radio, alerted local listeners that Three Mile Island was experiencing difficulties.[32] However, many people did not tune in to this report leaving many reliant on word of mouth; further, little information was known about the problem, and little was reported even when the Associated Press confirmed an accident shortly after the radio release. The reveal was not met with too much concern because many newspapers relied heavily on early statements from Metropolitan Edison and were quick to reassure the people that the event at Three Mile Island was not anything like *The China Syndrome*.[33] Harrisburg PA's *The Evening News* confidently comforted readers in a headline entitled 'Leak poses "no danger" to populace'.[34] On television, TV27 News opened with Lieutenant Governor William Scranton's reassurance that "everything is under control".

It was on March 29 that Metropolitan Edison held a press conference and offered reassuring words to the public. However, public attitudes were shifting by this point as papers such as *The Scranton Times* offered front-page headlines on problems, articulating growing concerns over the event and doubt over the positive statements of Metropolitan Edison. For *The Scranton Times*, the accident had become the "most serious" in American nuclear industry history.[35] Harrisburg's *The Patriot* did not deal with the incident in isolation and presented an article on the 'Legacy of Trouble' Three Mile Island had brought to the area, suggesting that while the crisis was unique, the plant had always been problematic.[36] The accident dominated the papers, with many articles bringing in stories relating to nuclear testing and past nuclear accidents as if to make the Pennsylvania crisis one of many nuclear perils.[37] Smaller articles ran on bomb threats, and some stories covered NRC meetings involving safeguarding against airplanes crashing into the plant.[38] Stories about terrorism suggested that worse could happen at the plant and that many dangers had not been divulged to the public by the authorities. This accident could well be one of many disasters, some seemed to imply.

While many members of the community continued as usual—shopping, socializing, entertaining, and working—others experienced significant anxiety as by March 30 front-page headlines read "Women, Kids Evacuated".[39] Although it was evident that this was "precautionary" (Thornburgh spoke of "an excess of caution"), the headlines speaking of evacuation would have come as a blow.[40] Three Mile Island was not only a concern in Pennsylvania—the event was one of national and international focus. An article in the *New York Times* spoke of a "wave of alarm" spreading across the country with heightened anxiety in remote states like South Carolina.[41] This fear was understandable considering that ABC News delivered a special news report speaking of the "nuclear nightmare" and the possibility of a meltdown.[42] Correspondent Tom Jarriel's statement that "thousands have packed their luggage and left, banks report many withdrawals, telephone lines have been busy" sounded like a narrative from apocalyptic fiction.[43] Jarriel also noted that many people were skeptical over official reassurances that the plant was not dangerous.[44] Likewise, *The Scranton Times* highlighted the lack of consensus among experts over the ramifications of the accident, with reporter Joseph X. Flannery directly commenting on the vagueness of statements by experts due to their own lack of information.[45] The inconsistences, generalizations, and confusion of official reports combined with intense media focus to considerably impact public anxiety and resentment, as we shall see in the next section.

Director of the Energy Management Agency in Eastern Pennsylvania, Robert Hetz, believed that the accident only made front-page headlines because little else of note was occurring at the time; he remarked that the event wouldn't have made a stir if it had coincided with a World Series.[46] Many news reports might be described as sensationalism as Luzerne County's director of emergency operations, Frank Townend, seemed to suggest.[47] One example of 'sensationalist' reporting arguably came from an article by the Associated Press (published in various papers including *Ocala Star-Banner*, *Wilmington Star-News*, and *Lawrence Journal World* on April 8, 1979), as the story was introduced with dramatic prose akin to a thriller novel. It read: "In the darkness before dawn, in the chill mists that rise from the Susquehanna River, the atomic powerhouse on Three Mile Island defied its human keepers and threatened catastrophe."[48] This article, tasked with detailing a timeline of events from March 28, also made reference to international protests, spoke of the sudden reality of *The China Syndrome*, and reminded readers of Hiroshima. Fear and anti-nuclear protests around the country were perhaps even exacerbated as the Associated Press circulated a map of the United States landmarked with Babcock and

Wilcox nuclear facilities (Babcock and Wilcox manufactured Three Mile Island). Other publications like Scranton's *The Sunday Times* reported nationwide panic on April 1, 1979 and reported protests in San Francisco, emerging lawsuits, emergency meetings in Nebraska, radiation checks in New York and West Virginia, and demonstrations in California.[49] On April 2, the Associated Press ran an article which appeared in many local papers (including *The Scranton Times*) stating demonstrations were occurring globally, even in Japan, over the inherently unsafe nature of nuclear power.[50] After the accident had passed, the incident continued to generate anxious stories; *Time Magazine* ran the feature article 'A Nuclear Nightmare' (April 9, 1979) and *The Bryan Times* described locals near the plant as the "human guinea pigs" (October 1979) of the nuclear power industry.[51] However, these reports should not be dismissed as 'sensationalist'. They were not articles that deliberately used language to incite intense emotion. It was not the intention for local papers to terrify or agitate their readers, which is something the official Kemeny Report deduced. Instead, reporters were grappling with a technological disaster of which they had little experience. Language and headlines exposed the fears, anxieties, anger, and confusion of the reporters who themselves were locals trying to make sense of a crisis. And we must remember that these reporters did not have access to the site and could not 'see' the event. They, just like the public whom their writing reached, were reliant upon conflicting sources. In the next section, we will look in more detail at how the crisis was conveyed to the public and why the event caused such a stir. World Series or not, and 'sensationalism' aside, the Three Mile Island partial meltdown did have a profound effect on public perceptions of the plant and nuclear power as a whole. Doubt over the trustfulness of the authorities, and inconsistences in reports created a communications crisis to rival the technological crisis occurring on the island.

Doubt and Inconsistencies

Initially, problems occurred due to the deficit of information, and when announcements were made they quickly became outdated. The wait for television news updates and the printed presses caused news 'down time' with citizens often floundering due to a lack of accurate and current understanding. The authors of *TMI 25 Years Later* make the point that the lack of thorough coverage "only served to heighten the drama".[52] Official statements, expert reports, and government announcements attempted to add clarity, but their conflicting accounts, lack of reliable and consistent data, and mistakes only confused and angered the public, with the media quick to expose inconsistencies and question the legitimacy of announcements. The NRC, Metropolitan Edison, and State announcements (often by Thornburgh or Scranton) usually conflicted. On March 28, Herbein argued that the accident did not appear to be serious, a claim corroborated by utility spokesman Dave Klucsik who declared "[t]here is absolutely no danger of a meltdown".[53] A series of carefully and strategically worded announcements by Metropolitan Edison dramatically minimized the seriousness of the situation; in fact, General Public Utilities confirmed that the seriousness of the crisis was miscommunicated by Metropolitan Edison. As noted by J. Samuel Walker, the claims issued by Metropolitan Edison—that significant levels of radiation had not been recorded—was "ambiguous and, since extremely high levels had been detected in the containment building, misleading".[54] Nevertheless, reassurances from the company led Scranton to declare "everything is under control". However, the situation worsened and the initial positivity expressed to the public shifted when Scranton altered his statement: "Metropolitan Edison is giving you and us conflicting information."[55]

Although Scranton still maintained that the risk to health was minimal he did state that, despite confused reports, radiation had been released. Further, the NRC's claims on March 29 that citizens were not in danger and that the risk was contained within the site were seen to be overly optimistic when vulnerable people were advised to evacuate the next day. While communication seemed to be failing between the plant, the NRC, and the State, the media and locals were becoming mixed up in the confusion and misdirection. This is partly why the President dispatched Denton to the site—he was to be one calm and clear voice on the matter and more helpfully candid. Nevertheless, word of the 'gas bubble' or 'hydrogen bubble' confused and worried the public even further—newspapers spoke of the potential for an explosion as a result. The word 'meltdown' was also used; it was "possible" the NRC said at a press conference.[56] Insurance companies started to field compensation claims, a curfew was introduced for those who had not evacuated the area, and although there were no reported cases of looting and violence, the Mayor of Middletown directed the police to act with lethal force if looting began.[57]

In an attempt to soothe the situation, on April 1 President Carter visited the area and nuclear plant with his wife and advised locals to remain calm. Carter's visit served to ease local fear and present the area as safe enough for such an important man to enter. Carter's background in nuclear physics also reassured the people that the President understood the situation, even if they did not. Yet, this did not entirely alleviate all concerns. In a special radio report on April 1, CBS News covered the President's visit to Three Mile Island, detailing the safety aspect of the visit and remarking, perhaps cynically, on the absence of Thornburg's pregnant wife and the

President's use of a radiation badge and radiation shoe protectors. On the same day as the visit, CBS's *Face the Nation* interviewed Senator Gary Hart on the crisis; this Sunday morning segment was introduced with the statement that the news arising from the plant was difficult to follow: "some of it seems contradictory; some of it is hard to understand".[58] Speaking of a possible meltdown and the hydrogen bubble, Hart, Chairman of the Subcommittee on Nuclear Regulation, said there was a risk, although hard to quantify, of a catastrophic accident and stated that if he lived close to the reactor he would move his family out if the situation worsened.

Although Thornburgh suggested the crisis was over on April 4, it was only when pregnant women and preschool children were advised to return on April 9 that the incident was considered over. However, the history of conflicting reports and retracted statements contributed to a sense that the locals had been lied to throughout the crisis. Allen Ertel commented on the apparent misdirection of the industry professionals: "I don't think anyone can say they told us the absolute truth."[59] A local resident who collected his child from school before evacuating told reporters that locals had been lied to and that the situation was graver than the authorities had admitted. Locals used terms like "panic", "distrust", and "cover up" and described their situation as "scary", prompting *The Sun* to declare that "trust was the biggest casualty".[60]

Poor communication was the cause. Richard Vollmer of the NRC explained that when he arrived at Three Mile Island information was hard to come by and the information shared with the press was often not qualified.[61] Herbein, a spokesman for Metropolitan Edison, commented on the similarities between the way he handled the press and the way the public relations officer in *The China Syndrome* dealt with the press. Herbein explained that when he watched the film he thought the public relations officer was incompetent but admitted "a week later I was doing the same damn things"; which, as James Weaver notes, suggests that even "capable and honest" persons can be forced to lie when "placed in the unfortunate position by the situation and by this technology".[62] And Herbein, not prepared for the role of spokesman, was in a tricky situation as information was demanded at a time when he simply did not know what was happening and felt pressured to make announcements. The situation was not just plagued with 'lies' but also with well-intentioned news distortion due to rumors, assumptions, and an early desire to be optimistic; Julius Duscha from *The Baltimore Sun* commented that third-hand information added to the web of misinformation.[63]

Inconsistencies in the press and reports had profound consequences and sparked voluntary evacuations, as one resident explained: "Everything was so conflicting in the news reports. You'd hear one (local) reporter saying there's nothing to worry about; you hear the national (news), the place is blowing up."[64] The 'Health-Related Behavioral Impact of the Three Mile Island Nuclear Incident' survey found that out of its respondents, 78% fled the area due to anxieties over conflicting information and subsequent distrust of authority and media information—a greater percentage than those who left for children (30%) and pregnancy (8%).[65] Conflicting reports from the industry and press cast considerable doubt over nuclear technology in general. Three Mile Island was quickly used to highlight other nuclear incidents that may have received little attention at the time; *The Globe and Mail* newspaper from Toronto, Canada made reference to fifty-seven nuclear incidents which shared similarities with Three Mile Island.[66] Even when Three Mile Island was relegated to the remote depths of newspapers, reporting did not entirely terminate, and talks of lawsuits, health studies, and protest movements were well documented.

In May 1979, Harrisburg's *The Patriot* delivered a powerful and emotive story about nuclear testing in an article juxtaposed with a small report on a negligence lawsuit against Babcock and Wilcox. This lawsuit entailed, as the paper pointed out, the recovery of damages for those believed to have suffered physically of psychologically.[67] This article was nestled in the middle of William Hines' 'Still Plenty of Fallout Over '50s A-Bomb Tests. When Science Prostituted, Everyone Loses'. Hines' article, which was introduced with the emotive imagery of child graves in St George, focused on health concerns following radiation exposure from Nevada nuclear tests. The report spoke of leukemia and a 'conspiracy' to cover up the health risks of nuclear fallout, which further delivered a damning blow to faith in the authorities and the technology. The editorial decision to place this piece, which explicitly referred to the misuse of science and its ill effects, next to the civil suit against Babcock and Wilcox was deliberate. Hines said of the prostituted nuclear industry: "the lady is still a tramp and her rehabilitation will not come easy"—indeed, for on this page both sides of the nuclear coin (the military and the peaceful atom) fell under critical scrutiny.[68] Even when the intense media attention declined, articles continued to question the industry and the events of March/April 1979. The *Tuscaloosa News* (October 16, 1979) reported on findings of the Kemeny Report (before it was officially released) and summarized that the report would show that industry at large was "run by people who don't know what they are doing—or don't care".[69] The Washington *Observer-Reporter*, January 25, 1980, ran the headline 'Three Mile Island Meltdown Was Near' and dealt with how the Rogovin Report found that meltdown had been close.[70] These concerns, fears, rumors, and anger over the crippling combination of industry failure, as

'Just' a Technological Crisis?

Allan Mazur describes Three Mile Island as a "melodramatic 'media event'" which dominated almost "40 per cent of the evening news on television networks during the first week".[71] The press was required to 'hit the ground running' when it came to conveying the technological complexity of Three Mile Island and was required to balance a melting pot of science, politics, technology, and personal accounts. In 1980, the Association for Education in Journalism met at Boston University on August 10 to discuss how, in the twentieth century, the media can respond to technological news stories in a meaningful and balanced way. The meeting, entitled 'When Experts Differ: The Role of Mass Media in Scientific and Technological Controversy', examined how emotive, technology based news stories responded to crises such as Three Mile Island. The special meeting highlighted the changing role of the media and the need for a new approach when 'traditional reporting' converges with scientific/technological news stories and political context.[72]

Three Mile Island was not just a 'technological' story, though. Mazur notes that other technology based accidents around the same time as Three Mile Island, like the crash of a commercial DC-10 airliner in 1979, received less media attention despite a number of fatalities.[73] Why was this? Arguably, it is because plane crashes involve limited casualties and a limited impact zone—although clearly a tragic event, there is a limit to how many people can be killed in a plane crash and a limit to how much environmental destruction can be caused. With a nuclear accident, the scope of human and environmental impact can be much bigger. Further, and most importantly, many people understand plane crashes and their ramifications but nuclear accidents are harder to comprehend. So, it is not simply the case that technological stories get special attention. Nor is it the case that all nuclear incidents receive intensive media coverage—Windscale in 1957 received much less media attention than Three Mile Island (even counting the resurgence in the 1980s) and the accident at Browns Ferry (when a fire started after a worker looked for air leaks with a candle) before Three Mile Island received less attention. Mazur explores why Three Mile Island received such instant and far-reaching press coverage. Mazur accredits Three Mile Island's notoriety to several factors that make it unique: the accident coincided with the Hollywood film *The China Syndrome*; the location of the plant made it easily accessible to reporters; it came at a time when President Carter was pushing the energy program; and, many reporters in print and television media seemed 'skeptical' over nuclear power.[74]

To this list, I would add several more reasons: Three Mile Island also occurred after the industry had failed to live up to promises and was already dwindling; that the accident occurred after a problematic 'Atoms for Peace' movement; that it was one prominent nuclear accident after many; that the American people were increasingly skeptical over the authorities (Watergate) and were consequently more vocal and prone to protest (Vietnam War protest). It could also be claimed that the date of 1979, just before a new decade, cast the technology as past its prime, failing, and outdated. As the 1980s approached and the human race crept closer to a new millennium, nuclear power in the future was under doubt, especially as the environmental movement spoke so positively about renewable energy. The fact Unit-2 had only been critical for a short time contributed to a feeling of chaos because soon after it started operating large issues had been uncovered. This was not a case of an old plant experiencing age related problems after years of faithful service; this was a new unit not long opened. If a new unit could become so dangerous, then what was to be said about the older plants scattered around America? In 1979, the people were left wondering if the 1980s had a place for nuclear power. Cleanup would stretch into the 1990s, meaning that the disaster of the present would become the problem of the future.

well as woefully poor communication, led to severe mistrust of the nuclear power industry, an issue Carl Walske (President of the Atomic Industrial Forum) noted was "virtually non-existent before the accident".[75]

REACTING TO THREE MILE ISLAND

Optimism, Fear, and Ambivalence

Despite vocalized concerns and anti-nuclear spikes, many citizens had faith in the plant, its engineers, and the government, and did not report notable distress during the crisis. This strong resilience was down to several factors already highlighted in this book. First, there remained strong optimism in society surrounding modern technology. As we saw in Chapter 2, the impressiveness of sublime technological developments (especially in the long 1960s) were things that inspired pride in many. The sophisticated nature of developments like nuclear technology and space flight suggested substantial expertise beyond the comprehension of the 'average' citizen, suggesting to many that the experts would have the situation in hand, even if the crisis was incomprehensible to a layperson. Scientists and engineers were accredited with winning the war and holding the Cold War at bay, so distrust at this stage over what seemed like a minor plant malfunction seemed melodramatic to some. The government had done well to sell nuclear power as the peaceful and benevolent atom and those who remained unaffected by the Three Mile Island crisis were those most likely to not associate the dangers of a nuclear plant with the catastrophe of nuclear weapons. Second, the local area surrounding the plant was no stranger to disruption and crisis. Those along the Susquehanna had dealt with natural disasters such as flooding and had also lived with the nuclear plant for several years; so, a 'slip-up' seemed a minor hitch in the otherwise secure operation of the plant. Most importantly, the plant was a vital source of income for many. Resident John Viselli stated that when the plant came to the Susquehanna River many were unconcerned because many benefitted from the development of the plant—this included everything from jobs to energy improvements.[76]

Some reports attempted to contextualize the crisis, or compare it to other threats, in an effort to defuse concern. In the *Wellsboro Gazette* (April 1979), an opinion piece made the point that traffic accidents might not be sensational but they are as lethal as nuclear disasters.[77] The *Pittsburgh Catholic* noted that alcohol and smoking posed threats to humans and would kill more people than the accident at Three Mile Island.[78] James J. Kilpatrick from Utah's *The Deseret News* asked for calm and composure while mocking the last remaining "prophets of Doomsday"—such as the outspoken activist Mr. Ralph Nader.[79] By concluding that nuclear safety was more robust following the Three Mile Island incident, he berated those like Nader who speak out against nuclear power when it is ostensibly a safe and clean option.[80] Alfred L. Peterson, a retired Army lieutenant-colonel writing in *The Baltimore Sun*, used the accident as a way to warn the public against what he considered to be the 'real' danger facing the community: nuclear explosives.[81]

The ambivalence that dominated the early industry also played a role here. On March 29, Roger Quigley from *The Patriot* commented on the ambivalence witnessed in Goldsboro, describing the mood as divided between "tranquility and anger".[82] Some witnesses mentioned that nuclear power was not a concern, while others noted that they were "scared to death".[83] In many ways, the Three Mile Island accident acted as a microcosm for the ambivalence noted in previous chapters about attitudes to nuclear technology in the wider community. Polar responses were common in many reports: some witnesses believed their experience of fear was a rational response to danger; others interpreted their lack of worry to be a logical response to a situation under control. For example, Anne Trunk, a Pennsylvanian mother of six, refused to evacuate, calling her response "calm and logical".[84] However, Trunk recalls her neighbors suggesting she was foolhardy in staying.

While some were ambivalent, or did not view the crisis as a big deal, others were flat out scared. Terminology and reference to valves, radiation, malfunctions, reactors, and so on presented the accident as an incomprehensible disaster. For some, the shadow of Hiroshima, as well as the controversy surrounding nuclear testing, made those strange buzzwords—especially radiation—seem apocalyptic. In fact, Plowshare had been rejected by the community partly because of worry over radiation. Despite being familiar with localized problems such as flooding, those anxious about the Three Mile Island accident lacked a comparable frame of reference because the reported accident had no familiarity. What does a nuclear meltdown look like? What happens? This was no flood. Furthermore, technology itself, while capable of extraordinary feats (such as putting an American on the Moon), was also responsible for catastrophic accidents. Scandals such as the Silkwood case and memories of authority failure and apparent misdirection during incidents such as the Centralia fire prevented some from having faith in the government and industry to solve the problem and present the public with the truth. *The China Syndrome* presented a popularized understanding—although fictional—of what could happen at Three Mile Island and the story was not positive.

Furthermore, the Three Mile Island event shook a sense of pre-existing ambivalence with some taking a negative stance on nuclear power after being previously neutral. Art Buchwald's opinion piece in *The Scranton Times* spoke of his ambivalence, pre-Three Mile Island, towards nuclear power: "I can go either way on nuclear energy, depending how the wind is blowing."[85] However, living close to Three Mile Island after the accident made him more concerned about the direction of the wind.[86] Even Walter Creitz, president of Metropolitan Edison, reflected on nuclear power ambivalence before the crisis and remarked that "maybe we should have been more pessimistic".[87]

For some, a nuclear power accident was "inevitable"[88] and Three Mile Island merely confirmed the idea that the technology was somehow uncontrollable and inherently dangerous; as chemist Chauncey Kepford proclaimed on April 1, the plant seemed "beyond human control".[89] With fears of an unsafe technology getting 'out of hand', rumors also started to circulate and worsen the situation. New Cumberland Council President Jack Murray said emotions were heightened with rumors contributing to fear.[90] Terror for many was apocalyptic in severity and some discussions about the event even relied on apocalyptic ideas—as seen with *The Catholic Witness*. This publication accused the government and industry as having suicidal mentalities and called for a reconsideration of technological usage alongside "moral theology" in order to avoid "self-destruction".[91] References were made to mass death and "genetic assault"—an apparent nod to the loss of life recorded in Hiroshima and Nagasaki. Further, apocalyptic undertones to the article expressed a rejuvenation of traditional and antiquated ideas of sin, judgment, and disaster with science and technology as a modern medium through which the human can be undermined, threatened, assaulted, and extinguished. Such sentiment echoes the analysis provided in Chapter 1, as championed by thinkers such as Roslyn Weaver, on the secular apocalypse. The Kentucky *Daily News* reported "grim" prayers in church, citing Goldsboro's Reverend Richard Deardorff who advised his congregation that the event was God's way of saying "be careful".[92] In an opinion piece in the *Pittsburgh Catholic*, the claim was made that nuclear power was against Christian values as the technology itself is a "Pandora's box" of many dangers.[93]

Radiation: What's Going On? Struggling to Understand the Situation

Reviewing *The China Syndrome* in March 1979, Vincent Canby questioned the plausibility of the film and whether the dangers articulated in the movie were accurate. Without technical training, Canby concluded that he must accept the plot's plausibility. He further noted that because he had experienced blackouts caused by machine and human malfunction, the film's theme of machine and human failure seemed credible.[94] Nuclear power is hard to understand and *The China Syndrome* raised questions that most viewers were not equipped to answer. If Canby was uncertain about the events in *The China Syndrome*, those trying to comprehend Three Mile Island weeks later would have a harder job.

Confusion and fear over radiation separated the Three Mile Island event from other technological crises. James H. Johnson and Donald J. Zeigler in *Economic Geographic* argue that radiological crises can provoke elevated levels of fear, distrust, and dread compared to other disasters partly because of the association nuclear power has with nuclear weapons.[95] Indeed, a poll conducted by the *New York Times* revealed that 36% believed that an accident in a power plant would produce a mushroom cloud (like the one that ravaged Hiroshima).[96] Even some experts were making this connection; for example, Ernest Sternglass, a professor at the University of Pittsburgh, claimed that the people faced 'fallout' equivalent to nuclear bomb fallout and said the community should "stand up and scream" over the radiation threats posed by the accident.[97] And certainly, there was panic. A child at the time, Kia Johnson recalls how parents raced into classrooms in desperation to remove their children,[98] and teacher John Milkovich recalls panicked parents collecting their children and covering their mouths against radiation.[99] Likewise, Dina Gonzalez, who was six during the accident, remembers how her mother instructed her to keep her mouth covered as she left school.[100] Layne Lebo recollects her horror as windows were closed and students were told to remain indoors,[101] while Tom Viselli remembers how his peers ran through the school in terror.[102]

Many anxieties surrounding radiation were down to conflicting reports over the seriousness of the releases and confusion about what radiation was.

An article in *Public Health Reports* commented that because most people did not have in-depth understandings of radiation, they were reliant on statements by experts; however, due to misinformation, contradictions, and retractions early on by experts regarding the event itself, when radiation came under discussion the experts were viewed with a measure of distrust based on a sense of apprehension towards expert opinions collectively.[103] Additionally, although many statements claimed that the public was not in direct danger from radiation, many of these statements differed, prompting General Utilities to state, with frustration, "the story has changed throughout the day".[104] *The Patriot* made reference to the enduring confusion which lasted beyond March 28, noting that while reports from the plant about radiation were "slow to come and confusing" the situation was not improved the next day with conflicting and confused reports continuing to surface.[105] This article highlighted the various conflicting information provided by State Police, the NRC, Metropolitan Edison, Pennsylvania Department of Environmental Resources, General Public Utilities, Lieutenant Governor William Scranton, Department of Energy, and Three Mile Island employees in relation to both the accident and the radiation release. Confusion only made the situation harder to understand, leaving many desperate for answers. On Sunday, April 1, 1979, during a sermon, Reverend Deardorff prayed to God for strength and enlightenment: "As we gather here this morning. It is with fear and trepidation in our hearts because we do not understand what is happening."[106]

However, anxieties over radiation were not just down to confusion and mistrust surrounding official statements; as argued by Marilyn K. Goldhaber and others, radiation anxiety arose because of the established idea, adopted by the people, that radiation was (and still is) "mysterious and dangerous".[107] The atom, as we saw in the last chapter, has always been linked to the mysterious and magical. However, during a time of crisis this mystery was less thrilling and more terrifying, especially as the 'magic' of radiation was linked to death, and the invisibility of radiation (exciting during the early days of X-rays) had become an insidious force. As such, common reactions to radiation mentioned its imperceptibility; as Jarriel for ABC news exclaimed, "people are afraid of what they can't see—radiation".[108] Radiation made this crisis different to any other experience as one resident noted "I've gone through fire, and I've gone through flood, (but) this radiation, you can't see it . . . and I guess that's why we (left)".[109] Similarly, Patty McCormick told *The Baltimore Sun* that she would have been less anxious if the crisis had involved a natural disaster because they are comprehensible; she wished for "something that I could see, that I could understand".[110] As the people could not rely on the experts, the crisis seemed even more chaotic and radiation became such an unknown terror that it was described by many as a sentient fiend. The lovable Nuclear Genie was now warped and threatening. Gill Smart remembers how his family "fled" to New York during the Three Mile Island incident; reflecting on his childhood memories he remarks: "I had the creeping sense of running away from a monster, from something seeping toward us like a spreading, metastasizing virus."[111] Even when scientific terms like 'rays' were used they were also described in dystopian terms: "The first thing I thought of is, are those rays coming out of Three Mile Island going to come to New York and harm my daughter."[112]

Many publications attempted to explain radiation in a logical way. On March 28, the front page of *The Scranton Times* reported that radiation had been released from Three Mile Island and under this a snippet article explained radiation in a piece called 'What is Radiation?' Such an addition highlights the newness of such concerns as the media found it necessary to add supplementary articles to explain facets of the news. In a very concise and clear article circulated by the Associated Press, radiation was explained as ionizing and non-ionizing with details on how radiation can manifest and when it might pose a risk.[113] Similar educational articles also appeared in Harrisburg's *The Patriot*. The front-page article on March 29 by Richard Roberts was heavily populated with facts and quotes from industry experts mainly concentrating on explaining the radiation leak. The lead story concluded with interviews with Three Mile Island workers who expressed no fear about the accident.[114] The second page of the paper attempted to explain how a reactor works and what a meltdown involves, complete with a small illustration of a containment structure. However, the same publication featured a small article on the Karen Silkwood trial, reporting that Silkwood's claims of defective fuel rods had been witnessed in X-rays by her supervisor Gerald Schrieber.[115] *The Patriot*, despite the educational section, was still delivering mixed messages.[116] Silkwood's trial was positioned underneath a photograph of Goldsboro citizens calmly shopping in town. On the same page, the article 'Plant Proponents See Fuel for Critics' offered an inconclusive debate considering whether previously reported problems at the plant were related to the current accident and whether the plant itself was inherently unsafe. This confused reporting continued on March 30, when *The Patriot*'s front page carried conflicting articles. The lead article '4 Counties Still on Alert' featured quotes from professionals including Pennsylvania State Health Secretary Gordon MacLeod who reassured the public that the radiation leak was not a threat, and Emergency Management Agency's Jim Cassidy who stated that it was unlikely an evacuation would be called for.[117] However, other articles on the front page included Peter J. Bernstein's '"Fowl-Up" Sketched for Panel', which considered the statement of NRC's chairman Joseph Hendrie who claimed the accident was a careless mistake; radiation was also a feature in this article with reassuring words from the NRC that radiation levels were not dangerous to the public.[118] However, this article was juxtaposed with an article by Mary Klaus bearing the more foreboding title 'Radiation Above Normal: Scientists Seek Closing'. Further, Klaus' article focused on the claims of Sternglass (University of Pittsburgh) and Dr. George Ward, a 1967 winner of a Noble Prize (in Physiology or Medicine), that Three Mile Island should never be reopened and that radiation levels were higher than authorities claimed.[119]

Confusion remained months after the incident as answers were still not forthcoming. Even outside the immediate geographical area, the confusion over the Pennsylvanian crisis drew attention; an article from Canada in Montreal's *The Gazette* (October 31, 1979) made this clear in its headline: 'Three Mile Island: Answers Are All "Maybes."' Resident Beverly Gorman also felt frustration over the 'maybes' and exclaimed "All I hear is rems and millirems. I never hear a question that is answered yes or no".[122] Her question was whether her children would be affected by radiation exposure resulting from the accident. Even when official reports claimed no radiation damage had been detected, rumors circulated about the apparent 'truth' of radiation contamination.[123] Chauncey Kepford from the Environmental Coalition on Nuclear Power (ECNP) spoke out at Pennsylvania State University and reported that his

THE ONGOING STORY OF KAREN SILKWOOD: SILKWOOD AND CRIME COMICS

Karen Silkwood's story appears in the first issue of *Corporate Crime Comics* (1977) in which the conspiracy theory surrounding her suspicious death is the focus: "What did Karen Silkwood know that we should know too?"[120] *Corporate Crime Comics* suggests that not only is the truth withheld from the people, but that any attempt to counter the industry's ingrained practice of lies and deceit may prove potentially lethal.

THE SILKWOOD TRIAL: RADIATION AND INDUSTRY CYNICISM

In March 1979, attorney Gerry Spence represented the Silkwood family in a lawsuit for Karen Silkwood's mental and physical suffering under Kerr-McGee. Spence's final words noting the excuses, lies, and shirked responsibility of Kerr-McGee would resonate with the discourse on Three Mile Island as questions were also asked about safety and trust: "They hid the facts, and they confused the facts, and they tried to confuse you, and they tried to cover it."[121]

radiation readings were higher than the NRC claimed,[124] and *The Patriot* ran a story on May 4, 1979 that higher radiation levels than assumed could be linked to additional cancer cases.[125] Stories included the mass death of wildlife from radiation poisoning[126] and there was a suggestion of mutated animal litters.[127] Some protestors publically discussed personal ailments like vomiting and diarrhea as symptomatic of radiation poisoning.[128] Robert Del Tredici's excellent book *The People of Three Mile Island* documents eyewitness accounts, interviews, and photographs, and contains many statements from residents noting ailments and unusual experiences as a result of the incident. William Whittock tasted metal when collecting the mail, Judith Johnsrud noticed an odor and experienced sickness, Jane Lee noted defoliation of trees, dairy farmer Clair Hoover reported cattle deaths, Henry and Nancy Gilbert described hundreds of bird deaths, Laurie Hardison explained that many of her chickens had died as well as goats, cats, rabbits, and sheep; in addition to Hardison's claims, Jane Lee further reported problems with horses, pigs, ducks, geese, and guinea pigs.[129]

Many claims made by residents, such as about cattle deaths, were countered by experts. On the subject of bird deaths, the director of the Harrisburg's Bureau of Radiation Protection, Thomas Gerusky, firmly stated that "[i]t sure wasn't radiation".[130] John Collins of the NRC also confidently declared that the death of animals from radiation contamination "does not exist" and public belief of contamination is due to not being informed on the issue.[131] Johnson and Zeigler suggest that anxieties over radiological crises were worsened because while the United States government optimistically celebrated nuclear power, they failed to discuss safety measures accurately; so the people did not know how to react which led to "overreacting".[132] During a Three Mile Island conference in 1980, Norman C. Rasmussen explained that radiation has been around for a long time and that information about it is readily available, but that most do not take an interest until there is a disaster. He noted that interest in radiation peaked when people were faced with the nuclear bomb.[133] This link between radiation and bomb fallout is one main reason fears also peaked during the Three Mile Island incident. Was radiation being leaked? Was it the same as fallout? Would the effects resemble those of a bomb? Further, the confused reports issued at the time prompted some people to question whether the experts were also having trouble understanding the situation. This is something Duscha from *The Baltimore Sun* commented on during the crisis, noting that muddled media reporting was a reflection of baffled experts.[134] It is of no surprise then, that in a letter to the editor in the *Wellsboro Gazette*, local reader Leonard Stuhler described the crisis as a "fiasco" and was convinced that the public had been negligently misled into blindly believing the technology was safe.[135] With suspicion and mistrust of the authorities still so high over this incident, it might be impossible to ever ascertain concretely, so that there is a consensus on what actually happened regarding radiation dosage, and as we will see in Chapter 5, the debate over the severity of the release continues into the twenty-first century.

Three Mile Island Fears: A Psychological Situation

Until the 1990s, the Pennsylvania Department of Health kept a record of over 30,000 citizens who lived in the five-mile radius of Three Mile Island during the 1979 accident. The department monitored the individuals for adverse health effects but ceased the program in 1997 after reports found no health problems in the registered group.[136] Findings indicated that there were no notable increases in cancer deaths or infant deaths linked to radiation from the accident; these findings were later supported in follow-up studies.[137] However, Barrett speaks of a "psychological situation" surrounding the reaction of the people who, although not panicking, "were very confused".[138] Arthur C. Upton, in his report on radiation health effects, argues that because radiation does not appear to be damaging, psychological distress and mental health have been perceived to be more serious.[139] In fact, the Kemeny Report conveys that "the major health effect of the accident was found to be mental stress".[140] As early as April 1979, newspapers were predicting "mental problems" and anticipated locals seeking treatment for stress and depression.[141] In the 1980s, State Secretary of Health, Macleod, noted a rise in people taking sleeping tablets and tranquilizers, and an increase in alcohol intake and smoking in his case study on public health and Three Mile Island.[142] Long after the incident, reflecting on the psychological toll the partial meltdown caused, Middletown Mayor Robert G. Reid worried about "flashbacks" if Three Mile Island reopened.[143]

The Three Mile Island incident became an important medium through which to study stress and psychological distress. A flurry of studies by different groups, using different methods and producing different results, dominated the months and years following the crisis. In fact, Steve Wing makes the point that there were more psychological studies of the population than radiation studies.[144] Initial studies included those by the Hershey Medical Center who sent out questionnaires, and Penn State University who conducted door-to-door interviews. The Pennsylvania Department of Health conducted telephone surveys from a random selection of the population, while the National Institute of Mental Health undertook a more focused project by concentrating on mothers of young children, Three Mile Island workers, and those with pre-existing registered mental health issues.

A critical study surfaced in April 1980 and was continued into 1981 by Peter S. Houts and others from Pennsylvania State University, College of Medicine, and The Pennsylvania Department of Health. The 'Health-Related Behavioral Impact of the Three Mile Island Nuclear Incident' report surveyed the population close to Three Mile Island. The team investigated the levels of distress experienced during the crisis, how stress levels were impacted by the events and the media, and the ramifications of stress on the health system and public policy. Through interviews during the crisis and three and nine months later, the data presented an overview of how locals responded (and continued to react) to the incident. This report found that the percentage of respondents registered as being "extremely" or "quite upset" about the 1979 incident dropped dramatically when the survey was taken again in 1980, demonstrating a peak of stress at the time of the incident (and most notably in those close to the plant).[145] Further, the number of those who reported concern that Three Mile Island posed a "serious threat" to the safety of their family declined from 1979 to 1980—again with highest figures recorded by those closest to the plant.[146] As expected, anxiety peaked at the time of the crisis and for those closest to the site of potential disaster. However, the psychological situation surrounding social and cultural reaction to the 1979 incident is more complex and we must be cautious not to assume that declining stress levels post-accident mark the end of psychological engagement. Three Mile Island provided a booster shot for nuclear fear, and left an indelible mark on the American psyche that would be visible in cultural products for years following the accident.

There is not enough space to examine every psychological study; but it is interesting to look at a few examples, especially those that were uniquely focused to gain an understanding of how nuanced psychological research is for this accident. One area on which a few studies focused was looking at how children responded to the crisis. A doctoral thesis written in 1982 focusing on the psychological impact of the disaster on children argues, somewhat controversially, that "the proposed and actual evacuation of young children out of Three Mile Island was similar to the traumatic event of World War II".[147] This study is not the first of its kind as it follows studies like the 1981 paper 'Children's Reactions to the Threat of Nuclear Plant Accidents' by Milton and Bernice Schwebel, which looks at elementary and secondary school age children from the Three Mile Island area who expressed anger, panic, fear, and resentment over nuclear technology. After the crisis, a vast majority of children, they found, expected a nuclear accident and expected there to be grave consequences as a result, with some articulating concerns over radiation dangers such as cancer.[148] In an interview by Del Tredici, Jane Lee explains that her eight-year-old nephew had such a severe reaction to the accident that he psychologically regressed to a three-year-old and became extremely emotionally distressed.[149] Interviews with children also found reports of chronic nightmares; many children also mentioned their confusion and how they prayed to God for help.[150]

Metropolitan Edison employees who were working at Three Mile Island at the time of the incident were also studied. The Three Mile Island workers were emotionally, economically, and occupationally involved in the crisis, and answerable to family, friends, neighbors, the press, and the wider industry. In 1982, an extensive telephone interview study was analyzed by Rupert F. Chisholm and Stanislav V. Kasl on the workers' response to the 1979 accident.[151] These studies covered male and female employees, parents, those with various levels of education, those with various years of employment at Three Mile Island, those who varied in regards to their economic situation, and those who were single and married. In many ways, the Three Mile Island workers represented a microcosm of the general population with the additional psychological complexity of being part of the nuclear industry. Surveys suggested that reactions to the accident were more acute (including higher levels of stress, anger, anxiety, etc.) for Three Mile Island workers in comparison to workers at other plants, such as Peach Bottom.[152] Such comparisons might suggest that stress levels are higher (or people are more susceptible to stress) the closer they are to the incident. The additional suggestion might be that the seriousness of Three Mile Island, while *articulated* elsewhere, was truly *felt* by the local community. The greater distance between the incident site and the respondent (for instance, the media) could account for a more ambivalent reaction. However, the fact that this was a nuclear power incident of national and international scrutiny would also have transferred anxieties related to the incident into the wider world. It is a generalization to suggest that those outside Pennsylvania did not experience elevated and persistent stress levels, especially if they were already vulnerable to concerns over this technology or lived near a power plant. It is too simplistic to suggest that those outside the area were not affected; on the contrary, numerous protests occurred across the country, including the Washington D.C. Rally on May 6 with record numbers of around seventy thousand protestors). As commented earlier, protests were even documented in Japan.

Concerns persisted beyond the Three Mile Island accident with peaks occurring at certain times, often coinciding with renewed media interest. For example, the anniversary of the Three Mile Island crisis saw a spike in stress as newspapers reflected on the 'near miss' a year earlier.[153] One major peak surrounded the reopening of Three Mile Island Unit-1 is confirmed by several studies. Unit-1 was shut down for refueling during the Unit-2 partial meltdown and the question of whether to restart the undamaged Unit-1 became a heated debate in the days, months, and years following the Unit-2 accident. In 1979, anti-nuclear group PANE (People Against Nuclear Energy) protested the potential reopening of Unit-1, citing the issue of increased severe stress in residents as a significant problem. The NRC questioned whether stress was indeed an issue so PANE took the issue to the United States Appeals Court for Washington District of Columbia (1982) arguing that under the National Environmental Policy Act the plant could be held responsible for causing psychological damage to locals. Initially, the court ruled that psychological stress should be treated as a form of pollutant and suggested that the negative impact on psychological health in locals should be viewed as an "environmental impact", especially in relation to the restart of Unit-1.[154] However, typical of most issues surrounding this crisis, even the issue of stress became subject to dispute and confusion. In 1983, the Supreme Court in *Metropolitan Edison Co. v People Against Nuclear Energy* intervened and stated that reported stress levels were based on a *perception* of risk posed by the plant and therefore stress could be caused by opposition to the nuclear policy in general and not linked solely to the accident. Consequently, the court decided there was no cause to seek termination of Unit-1's operation. The court was clear to state that this ruling did not mean the stress experienced and reported following the crisis was not serious and genuine. The problem lay with ascertaining whether the stress PANE reported was about "*attitudes toward* a facility" or the "*psychological impacts created by* that same facility".[155] This, the court found, was impossible to determine and so the ruling of the Appeals Court was overturned. Such a ruling is not surprising considering that this book has documented a complicated spectrum of anxiety and ambivalence towards nuclear technology since 1945. Was stress caused by *ideas* of the industry or solely by the *accident*? Either way, Unit-1 restarted in 1985 after years of debate about operational safety, trust, and local stress.[156]

Psychological studies continued years (and even decades) later, demonstrating the ongoing relevance of the Three Mile Island incident. In a study (1988) by Peter S. Houts, Paul D. Cleary, and Teh-Wei Hu, it is claimed that while short-term distress and damage were "substantial", the long-term effects of the accident varied in terms of the severity and enduring nature of the impact psychologically, socially, and economically.[157] Further, for others, Three Mile Island merely acted as a convenient 'excuse' for general issues facing locals; a letter from a reader in *Science News* spoke of blatant "scapegoating" by those researching stress at Three Mile Island and remarked that it is easy to blame general stress on this event.[158] Houts and others also make the point that there are discrepancies in some cases between the reality of the crisis and the perception of it, with some imagining worse scenarios due to various factors such as rumors and confusion.[159]

In 1990, Evelyn J. Bromet, David K. Parkinson, and Leslie O. Dunn revisited earlier research with refined parameters and looked at cases of depression, hostility, and anxiety rather than more general issues of 'distress'.[160] Like the earlier National Institute of Mental Health survey, this re-examination concentrated on mothers of young children within a ten-mile radius, Three Mile Island workers during the incident, and mental health patients within a ten-mile radius. Three years after the event, depression and anxiety were reported to be three times higher in mothers who believed Three Mile Island was dangerous compared to mothers who stated that they did not believe the plant was a risk.[161] Such reports led the researchers to conclude that the Three Mile Island accident had enduring psychological impact on the group.[162] Many reports saw a spike in depression and anxiety levels when Unit-1 restarted; some of these stress levels were higher than those recorded in the nine months after the 1979 accident.[163]

In 1993, M. A. Dew and Evelyn Bromet presented new findings of psychiatric distress over the Three Mile Island accident to *Social Psychiatry and Psychiatric Epidemiology*. Their study of 267 women who were living within a ten-mile radius of Three Mile Island during the accident examines levels of distress in mothers with young children, recording immediate psychological reactions and revisiting the participants over the following ten years. The report from Dew and Bromet builds on the substantial work completed in the fields of psychology and psychiatry relating to the incident.[164] Unlike other studies, this report maps ten years of psychiatric study on Three Mile Island. The results illustrate that the study group was split; one group displayed low levels of distress during the ten-year period of the study, but another group demonstrated higher levels of distress.[165] Dew and Bromet claim that those who experienced higher levels of distress had their distress levels rise and fall according to certain contextual events. While distress levels reportedly declined as years passed, the levels would peak when Three Mile Island returned to the news—such as with the restarting of Unit-1, as noted earlier.[166]

The numbers in the group registering low levels of distress were statistically greater than in the group that showed elevated distress levels; however, as Dew and Bromet explain, the minority group is by no means a *small* group: "while many residents were apparently never obviously distressed following the event, a significant minority experienced elevated symptomatology".[167] It is, therefore, important not to look at data and conclude that the Three Mile Island incident was not psychologically distressing because it did not involve a majority. In fact, the minority group was a statistically large group—far larger than expected—and the stress, for some, was sustained. It is also critical to consider the limitations of these psychological investigations, especially when examining what was called 'obvious distress', as distress can be masked and can manifest in less 'obvious' ways. Further, this investigation involved 267 mothers from a ten-mile radius and cannot be representative of the general public. However, studies show that a proportion of the public experienced notable symptomatic distress which was seen to peak at key Three Mile Island related developments. Such findings informed research on incidents like Chernobyl in which a catastrophe affected a local area. Dew and Bromet's investigation also assists psychiatrists in researching, over an extended period, how distress is subject to change—does distress remain stable or does it alter? The study also allows them to consider how distress can be profiled and predicted. Three Mile Island became an invaluable event through which to study psychological distress.

Part of Three Mile Island's unique psychological presence is due to the fact that the crisis was technological. One of the main differences between a natural and a technological crisis is that the technological crisis is often an 'accident' and thus is associated with blame and responsibility. Moreover, as Bromet, Parkinson, and Dunn note, technological disasters can have additional health ramifications often not associated with natural events: "Unlike natural disasters, the events at TMI unfolded over time, and have the potential for long-term threat to health."[168] On comparative psychology between Three Mile Island and other disasters, a study recorded in a special issue of *The Journal of Occupational Behaviour* notes how "the kinds of disasters typically studied—floods, hurricanes, explosions—differ distinctly from several psychological dimensions of the TMI accident" such as the clear boundaries of time and scale that natural accidents often have.[169] With natural disasters, most do not proceed to take a negative stance against weather, ecology, or geology. Three Mile Island was not just about a plant in Pennsylvania on the Susquehanna River; instead, Three Mile Island represented an industry and an entire technology. This is something the Rogovin Report notes:

> Through it all, the people around Three Mile Island have been models of admirable mass behavior under stress. They seem to accept, better than most, the fact of disaster, whether natural or manmade. . . . But they have about had it with this nuclear accident.[170]

Three Mile Island Fears: The Anti-Nuclear Resurgence

The radical publishing group *Fifth Estate*—a self-defined anarchist organization founded in 1965—issued a special publication of their periodical in April 1979 on the Three Mile Island incident and branded nuclear power as Frankenstein's awakened monster. In this special edition, activist Fredy Perlman's anti-nuclear article 'Progress and Nuclear Power: The Destruction of the Continent and Its Peoples' pits industry against nature, damning the "poisoning" of people and the environment in an event he describes neither as an accident nor surprising.[171] Perlman's perspective takes a hard line as he suggests that the mere presence of the plant represents civilized development that has historically tarnished the land since early settlement. In the same way as Native American settlers were misled and betrayed by authorities who represented 'civilization', so too have the American people been misled and betrayed by authorities who present nuclear power as something positive, he argues. The incident at Three Mile Island acts as an example, for Perlman, of the price of such 'civilization' and technological development. He suggests that the successful rhetoric surrounding the event, which describes it as being unforeseen and accidental, is intrinsically misleading.

Although the anti-nuclear movement surfaced in response to nuclear weapons, it experienced renewed focus and interest—if not an actual resurgence—when attention turned to nuclear power plants. A key moment of anti-nuclear power protest was experienced as a result of the Three Mile Island partial meltdown. As one protestor with the Newberry Township anti-Three Mile Island group claimed, to protest seemed "unpatriotic" until the Three Mile Island incident, which galvanized her to make a stand.[172] It was the Three Mile Island accident, as well as the crisis at Chernobyl, that reinforced pre-existing and general anti-nuclear criticism. It was also these moments that encouraged new protest participation and anti-nuclear sentiment. Nuclear activism enabled a sense of empowerment during a time in which participants felt the industry was perilous and out of control; it enabled those who opposed nuclear power to have a voice during a time in which they felt they were not being heard. Protesting also allowed members to feel engaged and proactive in an emotive and challenging debate that was historically left to politicians.

Some anti-nuclear protestors, especially activists, were already part of other movements (Civil Rights Movement, for example).[173] As such, many groups found additional layers of solidarity through the anti-nuclear movement. Feminism was one such movement. For Anti-Nuke Feminists in *Off Our Backs*, the Three Mile Island crisis was "the final straw" in growing anti-nuclear concerns.[174] In *Off Our Backs* (May 1979), the

Feminist Anti-Nuclear Task Force explained that the uniqueness of radiation hazards on women, fetuses, and children made nuclear power—and anti-nuclear sentiment—a feminist issue.[175] In fact, Emma Ogley-Oliver notes that motherhood for many acted as a "catalyst to their activism".[176] Some activists spoke of activism as a spiritual quest and part of a drive to protect Mother Earth; these aspirations were apparently linked to deep-rooted concerns over the environment and a sense of collective responsibility.[177]

Consequently, many anti-nuclear activists emerging in response to Three Mile Island were already experienced in activism and were able to rise up strongly to fill the void left by official miscommunication. With members of the public turning away from any authoritarian claim out of cynicism and mistrust, the activists—numerous in number and often with a tested history of proactive crusading—became, for many, a voice to listen to. Important and notable protest groups included the ECNP, PANE, and TMIA (Three Mile Island Alert); but also (although not limited to) Susquehanna Valley Alliance (SVA), and Anti-Nuclear Group Representing York (ANGRY). Although some were in existence before Three Mile Island (such as ECNP (1970) and Harrisburg's TMIA (1977)), the accident saw increased numbers for all pre-1979 organizations along with the establishment of aforementioned new groups such as the SVA, ANGRY, and PANE. Groups occupied many regions: Harrisburg (TMIA), Middletown (PANE), Newberry Township (Newberry Township Steering Committee), and Lancaster (SVA). A study published by Sherry Cable, Edward J. Walsh, and Rex H. Warland uncovered that following the accident, anti-nuclear groups formed 'coalitions' in which the groups divided their energies into two main units: the pursuit of nuclear education, and the pursuit of legal action.[178] Groups that united included PANE, ANGRY, and SVA under the ECNP; although attendance at ECNP meetings gradually declined from July 1979, the numbers experienced predictable peak turnouts in March, April, and July.[179]

Central to many protests was the demand that the plant not be reopened. Anti-nuclear protest was most keenly felt when residents, returning from evacuation, found that the cleanup process for Unit-2 would take many years at an astronomical cost.[180] The *New York Times*, on April 9, 1979 reported that over 1,000 protesters had gathered at Harrisburg to fight the reopening of the plant.[181] Even months after the event, nuclear power protests commanded exceptionally large numbers in Washington D.C. and New York. In May, petitions were handed out at polling stations in Londonderry during the primary elections; the petitions aimed to have Three Mile Island shut down permanently. In Middletown, protesters petitioned the council to seek the permanent

Figure 3.1 Anti-nuke rally in Harrisburg. Protester wears '2, 4, 6, 8 We Don't Want to Radiate!' sign at a protest rally in Harrisburg, 1979

President's Commission on the Accident at Three Mile Island, March 29–April 30 closure of Three Mile Island, and PANE achieved 1,500 signatures protesting the reopening.[182] PANE called for the cessation of nuclear power in the area and demanded a return to coal power. Clifford Jones, Secretary of the State Department of Environmental Resources, noted that anti-nuclear protest changed since Three Mile Island with many in Harrisburg openly questioning nuclear power when previously they were either pro-nuclear or ambivalent.[183]

The May 6 Washington D.C. rally included approximately 200 groups (such as peace groups, activist organizations, and religious associations)—with many coming from Harrisburg itself—all demonstrating distrust in, and rejection of, nuclear power. During the rally, activist Tom Hayden declared to the masses that the protest reminded him of the 1960s, referring to the heyday of the anti-nuke marches.[184] Jane Fonda, of *The China Syndrome* and known for her anti-nuclear stance, also addressed the people and proclaimed: "If we continue to place our health and safety in the hands of utility executives whose main goal in life is to maximize profits, then we will see more Harrisburgs".[185] Alongside the big names of Fonda and Hayden, the protest saw the return of icons Joni Mitchell, John Hal, Ralph Nader, and Governor Jerry Brown. At the rally, famous activist Dick Gregory went as far as to claim he would fast until American nuclear power stations were closed—Gregory had previously fasted in protest against the Vietnam War. As Luther J. Carter claims, the Washington rally showed the government that resistance to the nuclear industry was so strong as to be comparable to the power of anti-war and civil rights protests.[186] Three Mile Island was responsible for much of this outpouring. Indeed, Luther Carter claims that pre-Three Mile Island such a march would not have commanded the same numbers.[187] It was not only in May that unrest was being expressed. Fonda and Hayden remained part of the controversy for some time; they visited the Three Mile Island plant on September 24, 1979. Also, Nader spoke to mass crowds in New York City in September 1979 and exclaimed: "stopping atomic energy is fighting cancer. . . . Stopping atomic energy is saving this country".[188] Many protestors at the New York City demonstration held signs exclaiming "Stop Nuclear Power Before It Stops You!"

In many ways, the demand for collective consciousness and collective responsibility for nuclear power was central to nuclear protest. American architect Buckminster Fuller speaks of the tension surrounding the acceptance or rejection of nuclear energy at the time as "humanity oscillates" between concern over the risks associated with nuclear energy and the need for constant electricity.[189] Such a perspective might remind us of the ambivalence noted surrounding nuclear power in the previous chapter. However, for Fuller, the reason for such 'oscillation' is due to the public being "abysmally ignorant" of the technology.[190] Written a year after the Three Mile Island crisis, Fuller's critical perspective is tinged with both concern and exasperation as he urges people to consider the issues of atomic energy and finally make a decision on their stance rather than avoid the issue or fluctuate on it. Richard Pollock echoes these concerns when he speaks out over the idealism of the people towards nuclear power. Pollock was the Director of the Critical Mass Energy Project (CMEP). This project is a large anti-nuclear umbrella group in America founded by Nader who claims that the public would abandon nuclear plants in favor of candles if they fully knew and understood the information.[191] The CMEP has fought for the phasing out of nuclear energy and promoted alternate renewable energy sources. Although Pollock's perspective is biased, he highlights the concerns of the anti-nuclear movement and sheds light on the apparent problem with public indecision. Pollock speaks of two types of reactors: idealistic reactors and realistic ones. The idealistic reactors exist in a "technological paradise" in which nothing goes wrong and everything functions as intended.[192] The realistic reactor is the one that exists in a more problematized world in which industry can be susceptible to human error, technological malfunction, and accident.[193] The crisis at Three Mile Island was, for Pollock, an illustration of the realistic reactor: an example of what could go wrong. It was in the aftermath of the Three Mile Island accident that Pollock declared, "*Welcome* America, to the *atomic age*."[194] Pollock's suggestion is that while we have been speaking of the atomic age since 1945 with the bombing of Hiroshima and Nagasaki, America has only realized it with its first accident.

CONCLUSION: MAKING SENSE OF THE NONSENSICAL

While Three Mile Island did not signal the end of nuclear power, it did prove for Americans that nuclear power was not an entirely risk-free, peaceful technology despite what Atoms for Peace promised. As Thornburgh stated, "I don't think it 'tolls' the use of nuclear power, but showed that we can't run pell-mell into a form of energy we don't have a handle on."[195] The anxiety caused by the accident was arguably down to the unexpected revelation that the peaceful, benevolent atom could be potentially just as dystopian as the weapon being tested in the American deserts. Although the crisis did not spiral into a catastrophe, the accident proved that a potential catastrophe was possible. Chernobyl would make the point on that case.

As documented in the last two chapters, ambivalence was also experienced during this time; this is evident in psychological studies as well as in newspaper reports. As *The Baltimore Sun* noted: "Distrust and panic contrasted with a devil-may-care attitude."[196] However, while there were examples of both calmness and ambivalence, the incident did also cause considerable and debilitating stress for some. The shock and horror of the incident is very apparent in the language used to convey the crisis. In 1980, Clyde W. Burleson, in his historical text on nuclear accidents, referred to the "invisible hell which composes the innermost portion of a nuclear reactor core", and the "nuclear Medusa".[197] Even in news reports, attempting to (or claiming to) neutrally report on the accident, key words exposed anxieties. When interviewing a plant worker, ABC News (March 30, 1979) said "naturally he is worried"—the word 'naturally' signposting to viewers that worry is the logical and dominant response to the accident. In regards to the dangers posed by the plant, some publications referred to the "crippled plant" and spoke of the "human guinea pigs" living nearby.[198] Emotive language was also evident in how some locals expressed their feelings over the crisis with people referring to the plant as a "monster" and the technology as the "anti-Christ".[199]

Part of the problem, as we have seen, was that the experience of being adrift and having to rely on scraps of information disseminated by rumor, word of mouth, and contradictory articles contributed to a sense of powerlessness. Following the crisis, there was a concentrated drive to ensure locals felt that their feelings and opinions were heard. Speaking at the Eastern Pennsylvania Region of Hadassah Conference at Penn Harris Motor Inn in May 1979, Joyce S. Freeman argued that the Three Mile Island incident had fundamentally changed the way citizens viewed nuclear power, industry itself, and the role regulation plays. Speaking of vital changes needing to be made, Freeman noted that citizens "want to participate in a public discussion" not only surrounding the cleanup but in regards to any developments: "never again", she said, "will people be content with regulation after the fact".[200] Wanting to get involved, some residents took to opinion pages in local newspapers to air their views on nuclear power in a proactive manner; one interesting piece was offered by Ned M. Butt, a reader of *The Patriot* from Harrisburg, who outlined his ideas for improvements in the nuclear power industry. His ideas included design alterations focused on safety, higher trained experts in the plants, improved communication between all plant workers at all levels, the introduction of a new computerized system, and—importantly—public participation.[201] Butt's idea for public intervention even included public presence in the control room itself.

Wanting to be heard and wanting to feel like opinions can matter is vital during times of trauma in order for the individual or group to feel a sense of strength and even personal mastery over a very tumultuous situation. It is also a way for people to try to create a sense of coherency after a period of confusion and doubt. In the next chapter, we will start to see some ways in which popular culture attempted to make sense of the Three Mile Island accident. An opinion piece in *The Patriot* claimed that the Pennsylvania accident burst the "bubble of uncritical optimism".[202] It was through the arts, as well as critical work (academic writing, media publications, etc.), that the people would come to work through some of their feelings and ideas surrounding the bursting of that optimistic nuclear bubble.

Three Mile Island Souvenirs: Selling Disaster

Fiction can 'sell' disaster and present people with the ability to purchase event engagement through flights of the imagination. This exchange, rooted partly in the need to consume an event, can not only allow therapeutic exploration of an event but can also titillate those who enjoy revisiting disaster. The need to consume is also seen through the role of the souvenir. Here are a few Three Mile Island inspired products:

- Con-Cor produced an N Scale Nuclear Power Plant model kit with containment building and cooling tower. The 79-part kit proclaims on the front that it was modeled on the Three Mile Island plant. The constructed Con-Cor plant was designed to be used in conjunction with a model train as a background structure—a sign of what constitutes the modern landscape.
- Following the accident, many pieces of attire mentioned the crisis, often in satirical ways such as T-shirts proclaiming "Survivor: Three Mile Island Nuclear Accident. Caution. This Body Is Radioactive".
- 'Canned Radiation', which included a range of phony ingredients from 185 millirems of radiation to its own hydrogen gas bubble, went on sale as a gag gift. The 'Canned Radiation' tin pokes fun at the sensationalist claims of contamination by noting in small print its list of active ingredients as "baloney!" A little red warning on the can advises it be kept away from pregnant women and children but the product can be used, it suggests, as toothpaste for a "glowing" smile.[203]
- A postcard by United Souvenir & Apparel shows the plant foregrounded by the river. Despite the picturesque scene, the font spelling 'Three Mile Island' is a melting mix of purple, red and yellow set against a golden glow.

- Scouts in the Pennsylvania area have a Three Mile Island inspired commemorative patch marking 25 years since the crisis; the badge, featuring active, twin cooling towers, would have been purchased and attached to clothing.
- The restaurant Hooters still sells an extra hot sauce named '3 Mile Island'.

NOTES

1 'Harts Give Insights on Homesteading at Festival', *The Gazette*, Wellsboro, July 25, 1979, p. 10.
2 James Mahaffey, *Atomic Awakening* (New York: Pegasus Books, 2009), p. xvi.
3 Stan Benjamin, 'Three Mile Island and the Nuclear Power Industry', *Lewiston Journal*, March 17, 1980, p. 9.
4 Scott MacLeod, 'Life Not Back to Normal after Three Mile Island Mishap', *The Bryan Times*, October 1, 1979, p. 10.
5 Scott MacLeod, 'Life Not Back to Normal', p. 10.
6 Don Graff, 'Three Mile Island Revisited', *Daily News*, November 7, 1979, p. 10A.
7 Mahaffey, *Atomic Awakening*, p. xvi.
8 Scott MacLeod, 'Life Not Back to Normal', p. 10.
9 *The China Syndrome*, dir. by James Bridges (Columbia Pictures, 1979)
10 mc, 'the china syndrome', *Off Our Backs*, 9.5, May 1979, p. 3.
11 David Burnham, 'But Does it Satisfy Nuclear Experts?' *New York Times*, March 18, 1979, p. D1.
12 Thomas Tanner, '"The China Syndrome" as a Teaching Tool', *The Phi Delta Kappan*, 60.10 (June 1979), 708–712 (p. 708).
13 Ron Von Burg, 'The Cinematic Turn in Public Discussions of Science' (doctoral dissertation, University of Pittsburgh, 2005), p. 75.
14 Von Burg, 'The Cinematic Turn in Public Discussions of Science', p. 76.
15 Donald Jansen, 'Radiation Is Released in Accident at Nuclear Plant in Pennsylvania', *New York Times*, March 29, 1979, A1.
16 Quote from Gwyneth Cravens, in *Pandora's Promise*, dir. by Robert Stone (Impact Partners, 2013).
17 Reported in newspapers including: Barker Howland, '"The China Syndrome" Attendance is Up', *The Patriot*, March 30, 1979; 'Trouble at Ground Zero', *The Gettysburg Times*, 2 April 1979, p. 1.
18 Robert Del Tredici, *The People of Three Mile Island* (San Francisco: Sierra Club Books, 1980), p. 112.
19 *Our Friend the Atom*, dir. by Hamilton Luske (Walt Disney Productions, 1957).
20 For more on this see the United States Nuclear Regulatory Commission website which offers an animated diagram of PWRs and explains the process in easily accessible but detailed language. 'Pressurized Water Reactors', *United States Nuclear Regulatory Commission* (2015), available online at www.nrc.gov/reactors/pwrs.html (accessed July 15, 2016).
21 *Powering America*, dir. by Stephen Vidano (The Heritage Foundation, 2012).

22 Edward J. Walsh, *Democracy in the Shadows. Citizen Mobilization in the Wake of the Accident at Three Mile Island* (New York: Greenwood, 1988), p. 33.
23 'A Nuclear Nightmare', *Time*, April 9, 1979, 113.15, p. 8.
24 James A. Mahaffey, *Nuclear Power: The History of Nuclear Power* (New York: Facts on File, 2011), p. 128.
25 Colin Tucker, interview with Grace Halden, Sizewell B, April 13, 2015.
26 'A Nuclear Nightmare', p. 8.
27 Bonnie A. Osif, Anthony J. Baratta and Thomas W. Conkling, *TMI 25 Years Later: The Three Mile Island Power Plant Accident and Its Impact* (Pennsylvania: The Pennsylvania State University Press, 2004), p. xii.
28 See Del Tredici, *The People of Three Mile Island*, p. 11.
29 John G. Kemeny, *et al.*, *The Need for Change: The Legacy of TMI*, 'Report of the President's Commission on the Accident at Three Mile Island' (Washington D.C., October 1979), p. 8.
30 Daniel F. Ford, *Three Mile Island: Thirty Minutes to Meltdown* (New York: Viking Press, 1982), p. 44.
31 Mitchell Rogovin, *et al.*, *Volume 1. Three Mile Island. A Report to the Commissioners and to the Public*, Nuclear Regulatory Commission Special Inquiry Group (Washington D.C., 1980).
32 Francis T. DeAndrea, 'Evacuees Wonder if Life Will Ever Be the Same', *The Sunday Times, Scranton*, April 1, 1979, p. 1. Local resident Martha Viselli told *The Sunday Times* that in the immediate hours following the accident the people were not notified and she found out accidentally by radio.
33 For an example see 'Nuclear Plant Breakdown Releases Radiated Steam', *The Scranton Times*, March 28, 1979, p. 1.
34 Mary O. Bradley, Don Sarvey and Terry Williamson, 'Leak Poses "No Danger" to Populace', *The Evening News*, March 28, 1979, p. 1.
35 'Radiation Leaks from Nuclear Plant Continue', *The Scranton Times*, March 29, 1979, p. 1.
36 Barker Howland, 'Three Mile Island: Legacy of Trouble', *The Patriot*, March 29, 1979.
37 Mention made of Browns Ferry fire in 1975; radiation exposure at Trojan nuclear plant in 1978; and the explosion at Vermont Yankee nuclear plant in 1978, and so on.
38 See Anthony Cannella, 'Crash Data Sought by Nuclear Regulators', *The Scranton Times*, March 30, 1979, p. 1.
39 'Women, Kids Evacuated', *The Scranton Times*, March 30, 1979, p. 1.
40 'Women, Kids Evacuated', p. 1.
41 Wayne King, 'Concern Rises in South Carolina, Home of Many Nuclear Reactors; In Spite of Assurances and Plans for Safety Check, Worry Grows', *New York Times*, April 1, 1979, p. 30.
42 Tom Jarriel, 'Three Mile Island', ABC News, March 30, 1979.
43 Jarriel, 'Three Mile Island'.
44 Jarriel, 'Three Mile Island'.
45 Joseph X. Flannery, 'Some Experts Noncommittal', *The Scranton Times*, March 30, 1979, p. 1.

46 Hal Lacey, 'Nuclear Mishap Called "Minor" by Officials', *The Scranton Times*, March 30, 1979, p. 3.
47 Lacey, 'Nuclear Mishap Called "Minor" by Officials', p. 3.
48 Bob Dvorchak and Harry F. Rosenthal, 'Seven Days at Three Mile Island Nuclear Power Plant', *Daily News*, April 8, 1979, p. 6A.
49 See G. G. LaBelle, 'Nuclear Energy Concern Sweeps Nation', *The Sunday Times*, April 1, 1979, p. A-2.
50 See 'Demonstrators in Japan Urge Nuclear Shutdowns', *The Scranton Times*, April 2, 1979, p. 1. Demonstrations in 1979 were not always in relation to Three Mile Island, for example, demonstrations also occurred in Spain in early March 1979, before the American crisis, as a reaction against the growing Spanish nuclear industry.
51 Scott MacLeod, 'Life Not Back to Normal', p. 10.
52 Osif, Baratta and Conkling, *TMI 25 Years Later*, p. x.
53 'Pump Breaks Down. Radioactive Steam Escapes in the Air', *The Tuscaloosa News*, March 28, 1979, p. 1.
54 J. Samuel Walker, *Three Mile Island: A Nuclear Crisis in Historical Perspective* (London: University of California Press, 2004), p. 83.
55 Reported in Dvorchak and Rosenthal, 'Seven Days at Three Mile Island Nuclear Power Plant', p. 6A.
56 Noted in Walker, *Three Mile Island*, p. 144.
57 Reported in Dvorchak and Rosenthal, 'Seven Days at Three Mile Island Nuclear Power Plant', p. 6A.
58 *Face the Nation*, CBS, April 1, 1979.
59 'Lawmakers Privately Briefed at TMI', *The Patriot*, Harrisburg, March 30, 1979.
60 Antero Pietila, 'Distrust, Frustration Pervade Area Near A-plant as Some Prepare to Leave', *The Baltimore Sun*, March 31, 1979, p. A1.
61 Richard Vollmer, 'Representing the Nuclear Regulatory Commission', in *The Three Mile Island Nuclear Accident: Lessons and Implications*, ed. by Thomas H. Moss and David L. Sills (New York: The New York Academy of Sciences, 1981), pp. 110–113 (p. 110).
62 Morris K. Udall, *et al.*, *Accident at the Three Mile Island Nuclear Powerplant*, 'Oversight Hearing before the Subcommittee on Energy and the Environment of the Committee on Interior and Insular Affairs, House of Representatives', Ninety-Sixth Congress, First Session (Washington D.C., May 21, 24, 1979), Part II, pp. 6–7.
63 Julius Duscha, 'The Media and Three Mile Island', *The Baltimore Sun*, April 19, 1979, p. A19.
64 Cited in Peter S. Houts, *et al.*, *Health-Related Behavioral Impact of the Three Mile Island Nuclear Incident*, 'TMI Advisory Panel on Health Research Studies of The Pennsylvania Department of Health', The Pennsylvania State University, College of Medicine, and The Pennsylvania Department of Health (Pennsylvania: April 8, 1980), Part 1, Chapter 4, p. 4.
65 Cited in Houts, *et al.*, *Health-Related Behavioral Impact of the Three Mile Island Nuclear Incident*, Chapter 4, p. 5.
66 '57 Incidents Like Harrisburg Reported in U.S.', *The Globe and Mail*, April 17, 1979, p. 12.
67 'N-Plant Builder is Sued', *The Patriot*, May 1, 1979, p. 2.

68 William Hines, 'Still Plenty of Fallout Over '50s A-Bomb Tests. When Science Prostituted, Everyone Loses', *The Patriot*, May 1, 1979, p. 2.
69 Jack Anderson, 'Three Mile Island Facts Revealed. Nuclear Probe Shocking', *Tuscaloosa News*, October 16, 1979, p. 4.
70 'Three Mile Island Meltdown Was Near', *Observer-Reporter*, January 25, 1980, p. A5.
71 Allan Mazur, 'The Journalists and Technology: Reporting about Love Canal and Three Mile Island', *Minerva*, 22.1 (1984), 45–66 (p. 45).
72 'Conference: Reporting Scientific and Technological Controversy', *Science, Technology, & Human Values*, 5.32 (Summer, 1980), 34–35
73 Mazur, 'The Journalists and Technology', pp. 45–66 (p. 46).
74 Mazur, 'The Journalists and Technology', pp. 45–66 (pp. 59–60).
75 Carl Walske quoted in Benjamin, 'Three Mile Island and the Nuclear Power Industry', p. 9. Note: A study by Raymond L. Goldsteen, Karen Goldsteen, and John K. Schorr in 1992 showed that distress levels in locals during the crisis were related, in part, to fears over damage to health and feeling of powerlessness, in which mistrust of the authorities was one contributing factor. (Raymond L. Goldsteen, Karen Goldsteen and John K. Schorr, 'Trust and Its Relationship to Psychological Distress: The Case of Three Mile Island', *Political Psychology*, 13.4 (December 1992), 693–707).
76 DeAndrea, 'Evacuees Wonder if Life Will Ever Be the Same', p. 1.
77 'Uncommon Sense', *Wellsboro Gazette*, April 1, 1979, p. 2.
78 Jerome LeDoux, 'All in Perspective', *Pittsburgh Catholic*, July 6, 1979, p. 5.
79 James J. Kilpatrick, 'Three Mile Island—A Calm Perspective', *The Deseret News*, April 4, 1980, p A5.
80 Kilpatrick, 'Three Mile Island—A Calm Perspective', p A5.
81 Alfred L. Peterson, 'Three Mile Island and the Military', *The Baltimore Sun*, April 9, 1979, p. A12.
82 Roger Quigley, 'Goldsboro: Tranquility and Anger', *The Patriot*, March 29, 1979.
83 Ann Hartman quoted in: Quigley, 'Goldsboro: Tranquility and Anger'.
84 Ellen Hume, 'Three Mile Island Study: "Amazed" Housewife on Blue-Ribbon Panel', *Los Angeles Times*, April 12, 1979, p. A19.
85 Art Buchwald, 'Way Wind Is Blowing Shapes Nuclear Views', *The Scranton Times*, April 13, 1979, p. 6.
86 Buchwald, 'Way Wind Is Blowing Shapes Nuclear Views', p. 6.
87 'Not Prepared for Mishap, Operator of N-Plant Admits', *The Scranton Times*, April 13, 1979, p. 6.
88 'Nuclear Energy Fight Finally Goes Public', *The Scranton Times*, 3 April 1979, p. 6.
89 Anthony R. Cannella, 'Spurred by Nuclear Mishap, Foes Want Plant Stopped', *The Sunday Times, Scranton*, April 1, 1979, p. A-9.
90 Joseph Cress, 'Area Residents Remember Midstate During TMI Partial Meltdown', *The Sentinel* (March 23, 2014), available online at http://cumberlink.com/news/local/history/area-residents-remember-midstate-during-tmi-partial-meltdown/article_c0dd3214-b2e2-11e3-bed9-0019bb2963f4.html (accessed July 14, 2016).
91 'Nuclear Morality', *Pittsburgh Gazette*, April 6, 1979, p. 4.

92 Reported in Dvorchak and Rosenthal, 'Seven Days at Three Mile Island Nuclear Power Plant', p. 6A.
93 Marilyn Scranten Hunt, 'Nuclear Power a Pandora's box', The Forum, *Pittsburgh Catholic*, May 1, 1981, p. 5.
94 See Vincent Canby, 'China Syndrome is First-Rate Melodrama . . .', *New York Times*, March 18, 1979, p. D19.
95 James H. Johnson, Jr., and Donald J. Zeigler, 'Distinguishing Human Responses to Radiological Emergencies', *Economic Geography*, 59.4 (October 1983), 386–402 (p. 389).
96 Reported in John Chamberlain, 'Three Mile Island Perspective Lost', *Daily News*, Kentucky, May 2, 1979, p. 4A.
97 'A Nuclear Nightmare', p. 16.
98 Kia Johnson by email cited in: Diana Robinson, 'Where Were You When Three Mile Island Had Its Partial Meltdown? Readers Share Their Stories', *PennLive* (March 26, 2014), available online at www.pennlive.com/midstate/index.ssf/2014/03/where_were_you_when_three_mile.html (accessed April 12, 2015).
99 John Milkovich in an interview with Monica Von Dobeneck, 'TMI Stories: John Milkovich', *Pennlive* (March 22, 2009), available online at www.pennlive.com/specialprojects/index.ssf/2009/03/tmi_stories_john_milkovich.html (accessed July 19, 2016).
100 Dina Gonzalez in Diana Fishlock, 'TMI Stories: Dina Gonzalez, Scary Time for a 6-year-old', *The Patriot* (March 22, 2009), available online at www.pennlive.com/specialprojects/index.ssf/2009/03/tmi_stories_dina_gonzalez.html (accessed November 11, 2015).
101 Layne Lebo interviewed in 'TMI Anniversary Through the Eyes of Midstate Resident', *The Sentinel* (27 March 2014), available online at http://cumberlink.com/news/local/capital_region/tmi-anniversary-through-the-eyes-of-midstate-resident/article_23dd07dc-b616-11e3-a0a9-0019bb2963f4.html (accessed July 19, 2016).
102 Reported in DeAndrea, 'Evacuees Wonder if Life Will Ever Be the Same', p. 1.
103 Marilyn K. Goldhaber, *et al.*, 'The Three Mile Island Population Registry', *Public Health Reports*, 98.6 (November–December 1983), 603–609 (p. 603).
104 'Radiation: Who Said What', *The Patriot*, Harrisburg, March 30, 1979.
105 'Radiation: Who Said What'.
106 Quoted in 'Few Remaining Parishioners Think About "Living with Risk"', *The Scranton Times*, April 2, 1979, p. 1.
107 Marilyn K. Goldhaber, *et al.*, 'The Three Mile Island Population Registry', *Public Health Reports*, 98.6 (November–December 1983), 603–609 (p. 604).
108 Jarriel, 'Three Mile Island'.
109 Cited in Peter S. Houts, *et al.*, *Health-Related Behavioral Impact of the Three Mile Island Nuclear Incident*, Chapter 4, p. 1.
110 Pietila, 'Distrust, Frustration Pervade Area Near A-plant', p. A6.
111 Gil Smart, 'The Media and the Meltdown', *LancasterOnline* (March 20, 2011) available online at http://lancasteronline.com/opinion/the-media-and-the-meltdown/article_d78417a2-b901-5a16-b2f9-b714dfc6d64f.html (accessed July 19, 2016).
112 Quote from Gwyneth Cravens, in *Pandora's Promise.*
113 'What is Radiation?' *The Scranton Times*, March 28, 1979, p. 1.

114 '3-Mile Island Radiation Still Being Vented', *The Patriot*, March 29, 1979, p. 14.
115 'Deposition In Plutonium Case is Read', *The Patriot*, March 29, 1979, p. 8.
116 'Deposition In Plutonium Case is Read', *The Patriot*, March 29, 1979, p. 8.
117 Richard Roberts, '4 Counties Still on Alert', *The Patriot*, March 30, 1979, p. 1.
118 Peter J. Bernstein, '"Fowl-Up" Sketched for Panel', *The Patriot*, March 30, 1979, p. 1.
119 Mary Klaus, 'Radiation Above Normal: Scientists Seek Closing', *The Patriot*, March 30, 1979, p. 1.
120 The story is untitled and by R. Diggs. 'Untitled Karen Silkwood Comic by R. Diggs', *Corporate Crime Comics*, I (July 1977), pp. 2–6 (p. 6).
121 Cited in Michael S. Lief, H. Mitchell Caldwell, Benjamin Bycel, *Ladies and Gentlemen of the Jury: Greatest Closing Arguments in Modern Law* (New York: Scribner, 1998), p. 122.
122 Casey Bukro, 'Three Mile Island: Answers Are All "Maybes"', *The Gazette*, October 31, 1979, p. 17.
123 Three Mile Island's initial medical emergency plan was crafted by Ken Miller who noted that the hospital only treated one worker. Diana Fishlock, 'TMI Stories: Ken Miller, Wrote Medical Emergency Plan', *The Patriot* (March 22, 2009), available online at www.pennlive.com/specialprojects/index.ssf/2009/03/tmi_stories_ken_miller.html (accessed November 11, 2015).
124 Mary Klaus, 'Decision on Reporting is Political, Speaker Says. Air TMI Views, Public Urged', *The Patriot*, May 1, 1979.
125 'Radiation Dose Revised Upwards', *The Patriot*, May 4, 1979, p. 1.
126 Cress, 'Area Residents Remember Midstate During TMI Partial Meltdown'.
127 Edward J. Walsh, 'Resource Mobilization and Citizen Protest in Communities around Three Mile Island', *Social Problems*, 29.1 (October 1981), 1–21.
128 Walsh, 'Resource Mobilization and Citizen Protest', pp. 1–21.
129 Del Tredici, *The People of Three Mile Island*, pp. 13, 61, 23, 28, 64, 83, 87.
130 Del Tredici, *The People of Three Mile Island*, p. 98.
131 Del Tredici, *The People of Three Mile Island*, p. 118.
132 Johnson and Zeigler, 'Distinguishing Human Responses to Radiological Emergencies', pp. 386–402 (p. 389).
133 See Thomas H. Moss, 'Background of the Three Mile Island Nuclear Accident, I: General Discussion', in *The Three Mile Island Nuclear Accident: Lessons and Implications*, ed. by Moss and Sills, pp. 48–53 (p. 51).
134 Julius Duscha, 'The Media and Three Mile Island', *The Baltimore Sun*, April 19, 1979.
135 Leonard Stuhler, 'To the Editor', *Wellsboro Gazette*, April 18, 1979, p. 2.
136 Priscilla Dass-Brailsford, *A Practical Approach to Trauma: Empowering Interventions* (London: Sage, 2007), p. 266.
137 Evelyn O. Talbott, *et al.*, "Long-Term Follow-Up of the Residents of the Three Mile Island Accident Area: 1979–1988," *Environmental Health Perspectives* 111.3 (March 2003), 341–348. The official stance is to suggest radiation releases did not notably impact local health nor contribute to a spike in cancers or infant mortality. However, these claims are disputed by some groups as seen in Chapter 5.
138 *Powering America*, dir. by Stephen Vidano (The Heritage Foundation, 2012).

139 Arthur C. Upton, 'Heath Impact of the Three Mile Island Accident', in *The Three Mile Island Nuclear Accident: Lessons and Implications*, ed. by Moss and Sills, pp. 63–75 (p. 69).

140 John G. Kemeny, *et al.*, *The Need for Change*, p. 12.

141 'Mental Problems Are Expected for Residents in N-Plant Area', *The Scranton Times*, April 3, 1979. p. 1.

142 Gordon K. Macleod, 'Some Public Health Lessons from Three Mile Island: A Case Study in Chaos', *Ambio*, 10.1 (1981), 18–23 (p. 23).

143 'Three Mile Island Intensifies Nuclear Power Controversy', *The Hartford Courant*, September 28, 1979, p. W37.

144 There is a suggestion that the focus on stress over radiation is evident, perhaps, of a bias towards interpreting sickness symptoms to psychological problems. This will be explored in Chapter Five. Steve Wing, 'Objectivity and Ethics in Environmental Health Science', *Environmental Health Perspectives*, 111.14 (November 2003), 1809–1818 (p. 1810).

145 Cited in Houts, *et al.*, *Health-Related Behavioral Impact of the Three Mile Island Nuclear Incident*, Chapter 5, p. 6.

146 See the full report for an in-depth analysis of the data. For example, the report covers the variables and qualifications not dealt with here (such as initial sensitivity to stress in respondents, gender differences, and respondent bias—to name a few). Cited in Peter S. Houts, *et al.*, *Health-Related Behavioral Impact of the Three Mile Island Nuclear Incident*, 'TMI Advisory Panel On Health Research Studies of The Pennsylvania Department of Health', The Pennsylvania State University, College of Medicine, and The Pennsylvania Department of Health (Pennsylvania: April 8, 1980).

147 Penny Blagg-Miller, 'Unconscious Fantasies of Children Living in the Three Mile Island Area', Submitted in Partial Fulfillment of the Requirements for Degree of Doctor of Philosophy in The Union Graduate School (October 1982), p. 2.

148 See Milton Schwebel and Bernice Schwebel, 'Children's Reactions to the Threat of Nuclear Plant Accidents', *American Journal of Orthopsychiatry*, 51.26, 0–70. See also, Milton Schwebel, 'Effects of the Nuclear War Threat on Children and Teenagers: Implications for Professionals', *American Journal of Orthopsychiatry*, 52, 608–618. The work of Schwebel and Schwebel joins a wealth of research on child and adolescent psychological reactions to nuclear threat.

149 Del Tredici, *The People of Three Mile Island*, pp. 33.

150 Del Tredici, *The People of Three Mile Island*, pp. 101–102.

151 Among the results, this study illustrated that approximately one third of workers polled evacuated their family during the incident. Yet, as Chisholm and Kasl note, in context with the general population surveyed by other groups, results suggest that the families of Metropolitan Edison workers at Three Mile Island were less likely to evacuate. (Rupert F. Chisholm and Stanislav V. Kasl, 'The Effects of Work Site, Supervisory Status, and Job Function on Nuclear Workers' Responses to the TMI Accident', *Journal of Occupational Behavior*, 3.1, (January 1982), 39–62.

152 Chisholm and Kasl, 'The Effects of Work Site', pp. 39–62 (p. 56).

153 For example: 'Three Mile Island Meltdown Was Near', *Observer-Reporter*, January 25, 1980, p. A5.

154 Eliot Marshall, 'NRC Must Weigh Psychic Costs', *Science,* New Series, 216.4551 (June 11, 1982), 1203–1204. See also, *Metropolitan Edison Co. v People Against Nuclear Energy (PANE)* (1983).

155 William R. Freudenburg and Timothy R. Jones, 'Attitudes and Stress in the Presence of Technological Risk: A Test of the Supreme Court Hypothesis', *Social Forces*, 69.4 (June 1991), 1143–1168. Original italics.

156 Other lawsuits arose in response to the accident; another of note involved a settlement agreed by GPU in 1985 during which 280 residents sought damages for accident related injuries. Furthermore, in 1981, some money awarded in a class suit against the plant funneled into the Three Mile Island Public Health Fund. Osif, Baratta and Conkling, *TMI 25 Years Later*, pp. 87.

157 Peter S. Houts, Paul D. Cleary, and Teh-Wei Hu, *Three Mile Island Crisis. Psychological, Social, and Economical Impacts on the Surrounding Population* (London: Pennsylvania State University Press, 1988), p. 95.

158 J. Flack, 'Scapegoat for Stress', *Science News*, 127.18 (May 4, 1985), p. 286.

159 Houts, Cleary, and Hu, *Three Mile Island Crisis*, p. 97.

160 Evelyn J. Bromet, David K. Parkinson and Leslie O. Dunn, 'Long-term Mental Health Consequences of the Accident at Three Mile Island', *International Journal of Mental Health*, 19.2 (Summer 1990), 48–60 (p. 49).

161 Bromet, Parkinson and Dunn, 'Long-term Mental Health Consequences', pp. 48–60 (p. 55).

162 Bromet, Parkinson and Dunn, 'Long-term Mental Health Consequences', pp. 48–60 (p. 58).

163 Bromet, Parkinson and Dunn, 'Long-term Mental Health Consequences', pp. 48–60 (p. 56).

164 M. A. Dew and E. J. Bromet, 'Predictors of Temporal Patterns of Psychiatric Distress During 10 Years Following the Nuclear Accident at Three Mile Island', *Social Psychiatry and Psychiatric Epidemiology*, 28 (1993), 49–55.

165 Dew and Bromet, 'Predictors of Temporal Patterns of Psychiatric Distress' pp. 49–55 (p. 51).

166 Dew and Bromet, 'Predictors of Temporal Patterns of Psychiatric Distress' pp. 49–55 (p. 52).

167 Dew and Bromet, 'Predictors of Temporal Patterns of Psychiatric Distress' pp. 49–55 (p. 53).

168 Bromet, Parkinson and Dunn, 'Long-term Mental Health Consequences', pp. 48–60 (p. 57).

169 Chisholm and Kasl, 'The Effects of Work Site', pp. 39–62 (p. 40).

170 Mitchell Rogovin, *et al.*, *Volume 1. Three Mile Island. A Report to the Commissioners and to the Public*, p. 77.

171 Fredy Perlman, 'Progress and Nuclear Power: The Destruction of the Continent and Its Peoples', *Fifth Estate* (April 8, 1979), available online at https://archive.org/details/ProgressAndNuclearPowerTheDestructionOfTheContinentAndItsPeoples (accessed May 13, 2015).

172 Walsh, 'Resource Mobilization and Citizen Protest', pp. 1–21.

173 Walsh, 'Resource Mobilization and Citizen Protest', pp. 1–21.

174 Margie Crow, 'Nukes vs. Anti-Nukes: Malignant Monster Meets Critical Mass Movement', *Off Our Backs*, 9.5 (May 1979), pp. 2–6.

175 Carolyn J. Projansky, *et al.*, 'Nuclear Power is a Feminist Issue', *Off Our Backs*, 9.5 (May 1979), p. 5.
176 Emma Ogley-Oliver, 'Development of Activism: The Elders of the Anti-Nuclear Movement' (doctoral dissertation, Georgia State University, 2012), p. 68.
177 Ogley-Oliver, 'Development of Activism', p. 30.
178 Sherry Cable, Edward J. Walsh and Rex H. Warland, 'Differential Paths to Political Activism: Comparisons of Four Mobilization Processes after the Three Mile Island Accident', *Social Forces*, 66.4 (June 1988), 951–969 (p. 955).
179 Walsh, 'Resource Mobilization and Citizen Protest', pp. 1–21 (p. 11). Walsh notes that while many of these protest groups, especially the more dominant TMIA, PANE, SVA and ANGRY focused on Three Mile Island, ECNP also engaged in wider issues relating to nuclear technology in general—certainly pushing against the development and testing of nuclear weapons (p. 13).
180 Cable, Walsh and Warland, 'Differential Paths to Political Activism', pp. 951–969 (p. 953).
181 Alan Richman, '"Good News" at Three Mile Island, But 1,000 Stage Harrisburg Protest', *The New York Times*, April 9, 1979, p. D9.
182 Reported in Jon Harwood, 'Neighbors Vow To Bury N-Plant', *The Patriot*, May 22, 1979, p. 6.
183 Reported in Richard Roberts, 'TMI Startup "Blacklash" Seen', *The Patriot*, May 11, 1979, p. 57.
184 Luther J. Carter, 'The "Movement" Moves on to Antinuclear Protest', *Science*, 204.4394 (May 18, 1979), p. 715.
185 In *Pandora's Promise.*
186 Carter, 'The "Movement" Moves on to Antinuclear Protest', p. 715.
187 Carter, 'The "Movement" Moves on to Antinuclear Protest', p. 715.
188 In *Pandora's Promise.*
189 Buckminster Fuller, 'Introduction', in *Three Mile Island: Turning Point*, by Bill Keisling (Washington: Express Publications, 1980), p. 11.
190 Fuller, 'Introduction', in *Three Mile Island: Turning Point*, by Keisling, p. 11.
191 Reported in 'If The Public Knew . . . Nuclear Power Plants?', *Delaware County Daily Times* (September 4, 1973).
192 Richard Pollock, 'Foreword', in *Three Mile Island: Turning Point*, by Bill Keisling (Washington: Express Publications, 1980), pp. 13–15 (p. 14).
193 Pollock, 'Foreword', in *Three Mile Island: Turning Point*, by Keisling, pp. 13–15 (p. 14).
194 Pollock, 'Foreword', in *Three Mile Island: Turning Point*, by Keisling, pp. 13–15 (p. 14).
195 Quoted in Carmen Brutto, 'Nuclear Dangers Doubted', *The Patriot*, March 30, 1979.
196 Pietila, 'Distrust, Frustration Pervade Area Near A-plant', p. A1.
197 Clyde W. Burleson, The Day the Bomb Fell: True Stories of the Nuclear Age (London: Sphere, 1980), p.145.
198 'Three Mile Island Intensifies Nuclear Power Controversy', *The Hartford Courant*, September 28, 1979, p. W37.
199 Reported in Emily Yoffe, 'Pennsylvanians Lobby for N-Shutdown', *The Patriot*, May 8, 1979.

200 Constance Y. Branson, 'Legacy Seen in TMI: Public Demand to Be Fully Informed', *The Patriot*, May 16, 1979, p. 24.
201 Ned M. Butt, 'Nuclear Plants Can Be Made Safe', *The Patriot*, May 10, 1979, p. 62.
202 'Nuclear Protest: The Darker Side of Technology', *The Patriot*, May 8, 1979.
203 *The Original Canned Radiation*, Brenster Enterprises of Etters (Pennsylvania, c. 1980)

CHAPTER 4

Nuclear Reactions

Three Mile Island in Popular Culture

In 1979, just two months after the incident, psychology tutor Robert Coleman remarked that because residents struggled to understand the accident, they found it difficult to express their feelings over the event.[1] Here, we will explore how literature, film, and other cultural products produced following the accident work through many of the concerns and anxieties that we saw surface in Chapter 3. As we already know, the Three Mile Island accident occurred during the period James Bridges' *The China Syndrome* film was being screened in movie theatres, so the crisis has always been associated with narrative to some extent.

Due to the association Three Mile Island has with Bridges' film, and the fact that the accident became such a major televisual incident, it is also important to think about the relevance of the visual medium as well. The visual spectacle offered by film, television, and art coincides with the visual spectacle of the nuclear as linked to the iconic mushroom cloud. It is not a surprise then that the nuclear is closely linked to the televised. Further, as David Mark notes, the development of television after World War II in America and the evolution of nuclear development occurred almost simultaneously: "Television's growth and development as a mass medium, industry, and art form have occurred more or less simultaneously with the nuclear threat against all life as we know it."[2] Likewise, Joyce Nelson stresses that television aimed to exude technology mastery in much the same way as nuclear technology did: "television has been at the forefront of disseminating an ideology of technological omnipotence, the sign of which is surely the bomb itself".[3] The nuclear has always been very visual with test tourism, television, postcards, and representation in cartoons and comics. Also, for a long time television helped to normalize or soothe through presenting the nuclear in the 'living room' to families alongside their favorite television shows. Nelson argues that nuclear tests were televised not only to sell television sets in the fledgling industry but to dislodge images of nuclear disaster (cemented by Hiroshima) through showing "safe" tests.[4] Television helped to normalize the nuclear by sandwiching nuclear content between mainstream shows: "One minute you're watching, say, I Love Lucy; then it's The Life of Riley followed by the Yucca Flat Bomb-Test."[5]

However, considering that the chief anxiety with nuclear crisis is the invisible threat (radiation), the question of how this could be shown on screen, especially during news reporting, became a major issue. Arguably the problem of showing a nuclear power crisis is one reason why *The China Syndrome* does not explicitly deal with one. In the film, disaster is averted and a meltdown does not occur; in fact, the climatic moments in which the plant comes under threat last three minutes and the film's conclusion focuses on attempts by the journalists to release the truth to the public. There is no 'spectacle' in *The China Syndrome*. Televised news depends heavily on the hook of image, the brevity of story, the quick dissemination of information in a successive line of (often unconnected), stories and deals with snapshot images. What does this mean for the nuclear power industry accident that usually comes without the 'hook' image? What does this mean for portraying the invisibility of radiation? This was a problem faced by journalists reporting on the television news about Three Mile Island. With Three Mile Island most images were of the cooling towers, and this image has come to somewhat define the industry despite the vast amount of plants without cooling towers. At this juncture, the nuclear and televisual disconnect—visually there was little to televise. The nuclear, for once, had become 'invisible'. The Three Mile Island towers did little more than show viewers what the plant looked like on the day of calamity, but this was no different than how they looked the day before. Only the absence of clouds emitted from the stacks of Unit-2 suggested anything was wrong, but again, this was an absence rather than a presence. Instead, what seemed to materialize, in a rather ironic way, was the anxiety of being told there was a disaster but not being able to see this and understand it through image. What we had was a hidden threat and an invisible problem. Thus there were two types of response: some assumed that a lack of visual disaster meant no disaster at all, while others imagined apocalypse creeping closer in invisible clouds of radiation that no news correspondent could capture with film.

Partly due to a lack of visual stimuli during Three Mile Island reporting, the news reflected on earlier moments such as Windscale, the Urals, and Hiroshima and Nagasaki to help narrate and pictorialize the event. In a way, nuclear disaster is mnemonic—each emergency remembers previous crises. We might refer to 'hauntology' to explain such phenomena.[6]

Hauntology can refer to a 'ghostly' presence, what we might think of as a residual or repetitive moment. It also represents the 'spirit' of a moment; not just the facts and the historical context but the enduring essence of the event. Hauntology speaks of that which leaves a mark on culture long after the initial manifestation has passed—Hiroshima has certainly haunted all nuclear discourse and Three Mile Island would come to haunt (or influence) discussions on other accidents (especially Chernobyl and Fukushima as we will see in Chapter 5). Naturally, television news heavily relies on hauntology by referring to past events to help shape and contextualize the current story. However, in addition to this, other narrative techniques are used to help 'paint a picture' for an event which, like Three Mile Island, does not necessarily have a visual presence. Techniques familiar to creative writing were employed by the press to communicate an engaging and comprehensible story (see Chapter 3 for examples like the phrase "the darkness before dawn"[7]). Narrative techniques were also used to help bridge the gap between audience and disaster. During a time in which the people were seen to be at the mercy of conflicting information, many publications gave readers the ability to have their say. In May of 1979, *Life* magazine devoted an entire issue to the nuclear crisis with the headline "Judgment Day for Nuclear Power".[8] With a bleak and dark front cover image of Three Mile Island's cooling towers, the edition related public opinion on nuclear power—allowing the people to steer the story. The public, through text, would come to creatively participate in a situation they previously felt expelled and distanced from.

Fiction also helps to keep the story alive and, in many respects, has the proverbial last word. Ironically, as reporters described the Three Mile Island accident as "straight out of *The China Syndrome*",[9] fiction would come to present stories straight out of the Pennsylvania accident. The fascinating link between industry and fiction can be perfectly described through the following statistic: following the release of *The China Syndrome* and the Three Mile Island accident, shares rose for Columbia Pictures (the film's distributor) but shares in the nuclear industry dropped.[10] A tapestry is therefore woven in which fact and fiction become entwined in what we might broadly term 'nuclear narrative'. Indeed, all the elements for a good story are already in place—danger, innocent victims, and an invisible looming threat.

Fiction can make sense of a situation—especially one that is scientifically taxing or largely unprecedented. Lonna Malmsheimer, in her study of Three Mile Island and public response, notes that many people turned to imaginative works like Nevil Shute's post-apocalyptic novel *On the Beach* (1957) to help create an understanding of the situation.[11] More fittingly, *The China Syndrome* prepared people—in both positive and negative ways—for the realization of a nuclear power calamity. The film asked viewers how they would think, feel, and react if faced with a similar crisis. This is why, when the Three Mile Island incident occurred, some interviewees had already shaped their views in accordance with the film's themes. Through the influence of this movie we can see that visual protests in the street with billboards and vocal strength are not the only way concern can be registered. Literature (especially science fiction) is a medium through which people can express unrest over the present and project fear of the future.

In the documentary *Inviting Disaster*, the early morning period in which the accident occurred is described as "the low ebb of the soul" and a time in which "accidents happen".[12] The phrase 'low ebb of the soul' comes from the celebrated fiction writer Ray Bradbury. Bradbury's comment—as quoted in a serious documentary produced for the History Channel—shows the appropriateness of referencing fiction when contending with a major cultural moment. Often, fiction can be a way to explore trauma and consider the wider issues of a crisis. As Janet Kafka argues, the science fiction genre in particular deals with "any socio-political, ethical, or technological problem that the human race might meet".[13] Science fiction acts as the perfect medium through which people can explore scientific, philosophical, and social anxieties and this is possibly why the science fiction genre has been used not only to explore nuclear issues but also to work through some specific anxieties related to Three Mile Island and additional nuclear accident events like Chernobyl. The genre acts as "What If literature"[14] and allows authors to discuss future perils and ruminate over what could have happened if the event had panned out differently. *What if*, for example, the partial meltdown at Three Mile Island had been a complete meltdown? Fiction and news stories participate in an ongoing conversation with the public about pressing issues facing us today. The best thing about fiction, as opposed to most news stories, is that the public can produce this material themselves and, in that respect, have their voice heard.

The China Syndrome flags several dominant issues that Three Mile Island also articulated, issues that we will examine in this chapter: technology 'out of hand' through either equipment malfunction or engineering fault; concerns over dangers posed by nuclear technology (mainly radiation) and how this would impact locals and the local area; questions surrounding the trustworthiness of the industry, authorities and government; and the looming presence of the 'atomic cloud' as signaling impending potential apocalypse that, in this case, was meltdown rather than nuclear war. *The China Syndrome* was so crucial at the time because it also marked a concentrated shift from fiction focused on nuclear weaponry and war, to fiction that also critiqued the 'peaceful atom'. With Three Mile Island as a 'working example', more nuclear power fiction was produced after the 1979 event. Even though meltdown horror fiction experienced a boom with Chernobyl, the early origins are found along the Susquehanna River, a few miles from Harrisburg.

WRITING DANGER: DISASTER AND ITS EFFECTS ON THE COMMUNITY

Writing Radiation

In the previous chapter, many news reports covered the perspective that technology seemed out of hand and thus out of control. Many worries over technological malfunction during Three Mile Island echoed the warnings of *The China Syndrome*. Coming after a strong focus, especially through Atoms for Peace, on presenting nuclear power as magical, mystical, and miraculous, *The China Syndrome* tackles the spurious nature of such claims. In an early scene, Kimberly Wells (Jane Fonda) speaks of the plant's "magical transformation of matter into energy" before the plant succumbs to mechanical failure, human incompetence, and industry cover-ups.[16] The inherent unsafety of the technology in the film is implicitly referenced as engineers speak of malfunctions as 'routine'. Within twenty minutes of Kimberly speaking of the "magical transformation" that comes with nuclear power, there is utter chaos as technology fails and workers struggle to keep up with the unfolding series of breakdowns. The technicalities of the event are at times overwhelming, with rapid action and obscure terms like 'coolant' and 'relief valve' barked throughout the control room.[17] Many viewers, especially those without a working knowledge of nuclear plants, would know enough to appreciate that a terrible accident has occurred and that radiation is involved.

Radiation has long been an issue of contention for the nuclear age; in fiction, one of the most disturbing and enduring films

THE SIGNIFICANCE OF MASS CULTURE: HOLLYWOOD AND THE WHITEHOUSE

"I'd like to do this scene again", former actor Ronald Reagan joked after being shot in 1981. Between 1981 and 1989, Ronald Reagan was the 40th President of The United States. Interestingly, after starting his acting career in Hollywood, Reagan moved into the field of television and was the host for *General Electric Theatre*, touring plants delivering speeches for the company.[15]

dealing with radiation and its impact on the family unit was *On the Beach*. One harrowing scene featured parents choosing to murder their child and commit suicide in order to avoid the radiation sweeping across Australia. Families also attempted to escape radiation in *Panic in Year Zero!* (1962) by contemplating death through carbon monoxide poisoning. The trailer warned audiences that the film is shocking because "it can happen to you"; this message was likely to have resonated through the Three Mile Island disaster: nuclear crises once depicted at a distance could now happen to *you*.[18] Like many films of its generation, the biggest threat was not the weapon, not the explosion, and not the war itself, but the radiation that crept ever closer to family life. After Three Mile Island, radiation surfaced again as a dominant concern, mostly because it was both invisible and hard to understand. In her memoir entitled *Yes, I Glow in the Dark! One Mile from Three Mile Island to Fukushima and Beyond*, Libbe Halevy recalls her fright at being told to stay away from doors and windows during the crisis. Halevy remembers fearing a nuclear explosion at the plant and the way in which she grew increasingly concerned about radiation contamination: "unknowable horror threatened to launch itself at any second, I became transfixed by trying to pinpoint the exact moment the nuclear reactor would explode and take me with it".[19] The panic enshrined in Halevy's account is not unusual. As we saw in the last chapter, headlines during the incident and conflicting expert opinions only worsened anxiety over radiation.

Halevy's fears of a nuclear explosion at Three Mile Island decimating the area and obliterating her instantly never materialized; however, many fiction texts toy with the idea of a full meltdown and the ramifications of such an incident, such as Timothy Wentzell's *Faded Giant*. The title *Faded Giant* is derived from the code name used by the Department of Defense to refer to nuclear or radiological incidents.[20] The story imagines the scenario in which the partial meltdown at Three Mile Island was a full meltdown that was concealed from the public:

> We both sat there flabbergasted, realizing that what had happened was a full-core meltdown, and was not, as was over and over again told to the press and the rest of the country, merely an "excursion outside normal limits". This was something that could have led to catastrophic events far greater than any ever experienced even behind the Iron Curtain or, for the amount of radiation spread, Hiroshima.[21]

The main thrust of the novel involves engineer Harley Pelletier struggling to find justice within an industry that is described by Wentzell as corrupt and negligent. Not only did the authorities cover up a full meltdown, but health and safety, training, and working standards are described as criminally appalling. The training of Three Mile Island workers is portrayed by Wentzell as basic and flawed, and examination results for operators are suggested to be manipulated or faked. The consequences of these issues are reckless, negligent, and unskilled nuclear engineers. These themes are no doubt inspired by findings in the 1980s of cheating in operator examinations at Three Mile Island and the discovery that Metropolitan Edison had falsified information about leaks. Although Pelletier is a skilled nuclear engineer with years of experience, he is extremely paranoid over radiation contamination (not helped by the fact technician friends have cancer), and is suspicious over the safety and security of nuclear plants. The character of Pelletier is interestingly drawn as he has a predisposition to nuclear anxiety, suggestive of a more cynical twenty-first century protagonist. In *Faded Giant*, the 'expert' speaks and the words he speaks are filled with suspicion. Pelletier is French-Canadian and therefore apparently not seduced by the nuclear initiative so dominant in America. Arguably, Pelletier is the 'rational' engineer and expert who has not been 'contaminated' with the dreams and mysticisms of the Nuclear Genie; he is not burdened with the 'dream' and thus can face the 'nightmare'.

In 1982, Robert Lifton and Richard Falk spoke about what underpins the fear and mystery of the nuclear weapon. What they articulate is very relevant to fears associated with radiation and the nuclear power plant:

> This invisibility is part of the weapons' *mystery*. But the mystery also is importantly associated with our sense that we do not know, and cannot ever know, exactly what the weapons will do. . . . Hence the weapons are readily perceived as a kind of revenge of nature in the sense of possessing more-than-natural (supernatural) destructive power.[22]

In the discourse on nuclear power accidents, a similar sentiment is shared regarding the mystery and almost supernatural invisibility of the destructiveness of radiation. The concept of 'revenge', perhaps for negligence or foolishness (the reckless attempt to control the Nuclear Genie), is apparent in a range of fiction.[23] Lifton and Falk's nuclear "mystery" is linked to the idea of "tapping" a "destructive energy"; however, the *mystery* of nuclear power (presented as 'magic' through the Atoms for Peace project) is in *creating* energy that can potentially be destructive if things go wrong. The 'shock', in Alvin Toffler's sense (see Chapter 2), is to do with the fact we understand and expect "destructive energy" from nuclear weapons (it is what they were designed for). However, nuclear power is supposed to not pose a threat. This is something reflected on by Maryann DeLeo in the documentary *Chernobyl Heart* when she notes that the nuclear reactor seems so harmless but "there's nothing that shows the sort of violence of what came out of there because radiation is invisible".[24] Through DeLeo's work with Chernobyl children suffering from severe disabilities, she has seen the profound deformities and grievous medical complications caused by radiation exposure from the crippled plant; as such her mention of 'violence' is poignant. No bombs were dropped but the violence that befell these people, as articulated by DeLeo, is intense. Radiation alone can cast the *peaceful* plant as violent.

Writing on the Vulnerability of Family and Children

During Three Mile Island, many parents were naturally concerned about their children. Many reports described parents rushing to retrieve their children from school and covering their mouths. The dominant concern was the effect a radiation leak could have on pregnancies and babies. Even when, to soothe anxieties, Dick Thornburg visited the plant, the press noted that his pregnant wife did not attend. In the 1980s, following the accident, radiation continued to fall under investigation and continued to cause considerable anxiety for families.

'Violence' towards children by radiation in nuclear disaster fiction has always been a dominant theme. Arguably this is because children represent the future and the security of new generations; a threat to children is a threat to the sustainability of the human race. Previously, many nuclear disaster films focused on the war element or attacks on cities and countries. However, war based films soon made way for stories of nuclear catastrophe in rural areas. *Ladybug Ladybug* (1963) dedicates most of the film to a school class walking through the countryside to their homes after a nuclear alarm is raised. Set in the summer, the children walk in the sunshine down unmade rural roads, along tree-lined lanes, past sweeping fields farmed by tractors, past overgrown hedges, all the while discussing the ramifications of nuclear aftereffects. The area is very similar to rural Pennsylvania. Whether a bomb hits is uncertain (although heavily implied) but the threat to both the community and children is evident (in fact one child most certainly perishes after hiding from the attack in a refrigerator). Like the 1960s *Ladybug Ladybug*, the 1980 horror *The Children* focuses on youngsters in a picturesque suburban community returning home from school—in this situation, the children are injured by a nuclear plant release which smothers their school bus.

The Children is an excellent example of a dystopian nuclear power film that conjures associations with Three Mile Island (even without mentioning the accident itself) due to the film's timely release and comparable themes of mistrust, negligence, and radiation anxiety. Set after an accident in a nuclear power plant, *The Children* provides an extreme and gruesome portrayal of radiation-induced mutation as the youth of the town are transformed into radioactive zombie killers. The film is an extreme, and almost comical, example, but it was clearly inspired by the reported safe releases at Three Mile Island a year before. In *The Children*, a fictitious New England community becomes unknowingly contaminated by what are perceived to be normal emissions. Introduced with footage of a live power plant, the horror quickly progresses as a steam cloud is released from the plant and infects the local New England area. Due to the isolated nature of the town, the tragedy in *The Children* mostly remains secret—the absence of an explosion or plant alarm renders the crisis silent. The local news media believes the shutdown of the nuclear plant after the cloud emission is separate to the mysterious disappearance of the local children; this lack of correlation can be said to chime with claims by activist groups during Three Mile Island that plausible links were not being made between health concerns and the partial meltdown.

The children in the town vanish after being saturated by the leaked radioactive cloud, a reference perhaps to the fact that it is the children who are particularly vulnerable during a nuclear crisis. Children, it is implied, are the ones who will suffer from the arrogance of adults who allow such dangerous technology into the community. In a vengeful act, the children pursue and murder adults by masking their evil nature behind innocent, cherubim facades. The adults, fooled into believing the innocence of the children, are gruesomely killed leaving behind flesh stripped bodies. The infected children murder through flesh to flesh contact because their little bodies emit lethal doses of radiation. The touch based murders act as a metaphor for the insidiousness of radiation as innocent children become a 'carrier' for contamination: the children are silent and unsuspected but lethal. Furthermore, *The Children* uses the metaphor of infant killers as a commentary on the innocent facade of power plants, which are secretly and silently deadly.

The infected children can only be killed when their hands are cut off, and this might be a reference to Deuteronomy 17 from the Bible, where the hands of idolaters are removed to purge evil. The idolatry here is for the nuclear industry, and the cost of this idolatry is the elimination of future generations. There is only one God, Deuteronomy argues, and the film makes the case that it is unwise and deadly to put faith in the godly Nuclear Genie. The film argues that nuclear power is no God and anyone worshipping this power will be punished. There is also a strong emphasis in Deuteronomy on the parental role and the responsibility of families for their children; this is relevant when we consider that the dominant theme in *The Children* is the sacrifice of the younger generations in the pursuit of the 'abomination' of contaminating nuclear technology. At the end of the film, the birth of a contaminated murderous baby suggests the damnation of future generations born into the nuclear age. The adults are punished for promoting nuclear power at the risk of their family.

A more recent offering debuted in 2014 that explicitly features Three Mile Island is *My Father, the Old Horse* directed by Max Einhorn. This award winning, locally produced film deals with a family crisis during the Three Mile Island incident. The nuclear peril features as a backdrop to the more immediate familial issues facing a grandfather, father, and son who embark on a hunting trip near the island. While the family attempt to rescue their decaying relationship, the wider community is trying to navigate and survive the confusing time of the 1979 accident; confusion, sadness, and desperation saturate the film as the secrets and miscommunications of the family unit mirror the same issues in the wider community. While the accident features as a subplot, there are many nods to crisis throughout that are dark and poignant. During the hunting trip, the family encounters a series of disturbing phenomena such as secretive men in protective clothing sweeping the area with Geiger counters. Dramatic irony works well here because the viewer in the twenty-first century will understand the references to the disaster even if the family is yet to realize the severity of the situation. Intermittently, a broken radio feed issues updates on the crisis with announcements shifting from early reassurances to an eventual urge for evacuation. The conclusion features real sound bites reassuring the people that while radiation has been released, it is minimal and not a danger to health—an announcement the film contradicts minutes earlier when dead birds fall from the sky. As viewers, we are reminded that the Three Mile Island crisis involved families with complex lives and even when other issues (like family matters) take the foreground, concerns over radiation, untrustworthy authorities, and the malfunctioning machine lurk in the background. The film was shown during the thirty-fifth anniversary of the Three Mile Island accident along with *The China Syndrome*, further acknowledging that the crisis still lingers decades later.[25]

Writing on the Environment

The China Syndrome, *Yes, I Glow in the Dark!*, *Faded Giant*, *The Children*, and *My Father, the Old Horse* all deal with the rural community to various extents. Nuclear threat, namely radiation in these examples, is seen as perilous to families and future generations. However, environmental focus also positions the local landscape as under threat. In an early scene from *The China Syndrome*, a bird's eye view follows Kimberly Wells and Richard Adams as they travel through America. The long shot exposes the audience to beautiful rural America that will soon be subjected to a grave accident.

The 1980s saw a surge in environmentalism, particularly since the period was marked by Bhopal (a catastrophic accident in India in which hundreds perished when toxic gas from a chemical factory escaped into

Turning Nature into Technology

In Chapter 3, we saw how Glen Hart, during the Whitneyville Homesteaders Festival, commented that technology seemed to be controlling the people and changing the nature of the relationship between land and humankind. In his philosophical works, Martin Heidegger conveys the difficult and multifaceted relationship humankind has with modern technology. In a series of four lectures published collectively in *The Question Concerning Technology and Other Essays* (1977), Heidegger speaks of how technology can convert nature into a "standing reserve".[26]

How does this impact how we view Three Mile Island? Heidegger uses a hydroelectric plant as an example of his thoughts on 'standing-reserve', but this is also helpful to understand the relationship between the nuclear plant and the environment. As Liam Sprod in *Nuclear Futurism* summarizes, the standing reserve "means that the world is seen merely as a resource".[27] For us, this means that the river is viewed as a resource for cooling the reactor. Heidegger notes that the hydroelectric plant is perceived as under control of the human operator but in reality, the plant, by damming the river and converting the water into a component of the plant's functions, has actually taken control of nature and converted it into something different than what it was before.

So, a Heideggerian reading of Three Mile Island might suggest that the plant imposed upon the Susquehanna, assimilating it into part of the plant and transforming it from a basic natural watercourse within a hydrological cycle. The Susquehanna has been altered, interfered with, and harnessed as part of an artificial production. When Heidegger notes that there is danger in viewing this technological interference as natural, he partly means that there is a danger in looking at that water source and not seeing that it has been altered, and looking at the plant and seeing a disconnection between the workings of the plant and the impact this has on the environment. Heidegger would argue that since construction of the plant began, the environment had been imposed upon and transformed; while we see a structure jutting out of the rich local foliage, we do not realize how the whole local area had been converted into a 'standing reserve' to keep the Nuclear Genie fed and alive.

the community in 1984) and Chernobyl. These mass disasters came just after Love Canal (1978 environmental disaster in which toxic waste buried in the 1940s and 1950s started to affect local communities built on the site) and Three Mile Island. Undoubtedly, industry had changed the landscape and environment, and posed certain threats to the natural world.

In much environmental criticism, especially aimed at industry, the focus remained on visual disaster. Mutation, of the planet and the human, became a dominant theme through which to portray the 'invisibility' of radiation visually. In the graphic narrative *Trashman* by Manuel "Spain" Rodriguez (1968–1985), the character Trashman pushes against mainstream superheroes of the time after the polluted nuclear post-apocalyptic world imbues him with special powers. These comic books suggest that a post-apocalyptic nuclear world will not feature the standard mainstream heroes, but trash based ones. A nuclear world is trash—even the heroes in such a world will be trash. Another example of 'trashed' genetics is evident in Adam Scott Clark's dark comedy *Beaver Pig* which actively employs the Three Mile Island accident as a metaphoric moment of transformation from human to beast. The—as yet unpublished—manuscript, was a leading contender in Amazon's 2013 Breakthrough Novel competition. Out of 10,000 entries, *Beaver Pig* made it to the quarterfinals.

Clark makes his views on Three Mile Island very clear in the opening section of his story:

> From the heart of the island, buzzing high tension power lines span the river and drape over the cornfields while carrying the message of nuclear salvation and preaching the plant's power to unsuspecting subjects many miles away, where the sermon will eventually be but a mere whisper, a constant evangelical mumble in the ear, applying slight but perpetual psychological pressure to and manipulating the subconscious minds of the unwitting power consumers throughout Central Pennsylvania.[28]

The ominous presence of the plant permeates the air around it with an electronic 'buzz', and conditions and subdues the people who are both reliant on the plant and overshadowed by the dangers it represents. The link between energy and religion draws attention to the importance placed on power (especially following the energy crisis). There is a suggestion of a secular replacement of the traditional Christian God in American thought which corresponds to the godly Nuclear Genie in early nuclear propaganda. The fact that the danger of the plant is overlooked is symptomatic of the importance placed on energy over safety: "Herein lies the dual promise of infinite and transformative power and utter, almost whimsical, self-destruction, that glistening cusp where we as a species prefer to thrive."[29]

Are the people truly "unsuspecting" though? Clark suggests otherwise. The 1979 accident haunts the text and births not only Beaver Pig but a suspicious and angry town. The people are not unsuspecting at all; instead, they willfully ignore the fact that the power plant has mutated human existence in a multitude of ways since construction. This ambivalence is symptomatic of the atomic age and is represented here, as the characters in the novel simply do not want to face the truth of what the plant represents in this story.

Beaver Pig becomes an animal/human trans-species following the 1979 accident in which a radioactive beaver bite contaminates a young boy. The hybrid creature is initially described as an ugly beast, "an abomination of the natural world".[30] Beaver Pig invades homes and shops and is rumored to pose a risk to women and children (just like radiation was feared to do). The truth, however, is that Beaver Pig is a victim of the Three Mile Island accident. Calling on the nuclear-enhanced heroes of comic giants like Marvel, Beaver Pig not only represents the "monster inside everyone"[31] but comments on a changed world in which heroes and anti-heroes are not 'home grown' but genetically warped by one of the most concerning developments of our time. Like comic book heroes Trashman, Captain Atom, The Hulk, Dr. Solar, and Firestorm, Beaver Pig is a product of the nuclear age and a monument to lingering concerns and the imaginative preoccupation with the technological sublimity of nuclear power. Yet, the text isn't an entirely serious depiction. Clark's text comments on sensationalist rumors of genetic mutations and responds satirically to stories of 'cover-ups' and conspiracies after Three Mile Island:

> Despite government denial of the environmental impact, over the years, there have been reports made of birth defects and strange mutations in people, wildlife, and livestock in the region, all attributed to the radiation leaked from the meltdown. It was all hushed up but I know the folks who owned those reactors were sued for willful neglect, bad maintenance, and after denying it all over the place paid off something like 82 million dollars in partial settlement for class action suits and claims, including those damaged by the leak immediately and those parents who later had kids with birth defects.[32]

Beaver Pig, while satirical in places, is a metaphor responding to how some envisaged a hidden threat lurking in the community. Beaver Pig represents the threat of radiation for not only is he significantly mutated, but he insidiously hides within the community. His hybrid form also speaks to the concerns noted previously about environmental pollution. Throughout the text, the plant is depicted looming over the town and stretching far into the community: "Power lines span the river and drape over the cornfields." While the plant looms over the town, Beaver Pig lurks somewhere in the wild and, like nuclear testing and chemical accidents, the local environment is described as fouled as a result. A former idyll, crowned by the beauty of the Susquehanna River, is now a place of contagion.

Yet, for all the negativity he represents, there is a likeability to Beaver Pig. The creature's sharp wit, friendly demeanor, and good nature complicate what could have been a straightforward story about a demonic radiation soaked monster terrorizing the populace (like in *The Children*). The story's villain, Clark suggests, is the authorities; the monster is nuclear power and the reckless, selfish actions of the people who created Beaver Pig and who now mercilessly pursue him. Yet, Beaver Pig is presumed to be evil and is hated because he is so visually nonhuman and a living monument of the 1979 crisis that many would prefer to forget. The legend of Beaver Pig becomes a means of distraction from the real danger looming so close in the form of cooling towers: "tales of Beaver Pig were reiterated around campfires. Doors were locked and dead-bolted at night; lights were left on".[33] It is psychologically easier to deal with the tangible threat of Beaver Pig—an entity that can be locked out and killed—than it is to deal with radiation waves and water contamination that cannot be detected by putting on a light. Beaver Pig was never the threat but, distracted by the hunt, locals miss the root of all of their problems; namely, the plant that is regularly referenced looming in the background.

Mutation is not just the stuff of fiction, though; Three Mile Island's connection to 'mutation' is also depicted in Lauren Redniss' illustrated history text *Radioactive: Marie & Pierre Curie* (subheaded 'A tale of love and fallout'). Briefly abandoning graphic form for a page of photographs of flowers, Redniss shows a single rose that is malformed and without reproductive function alongside an overview of the 1979 accident.[34] The mutant rose serves to highlight the environmental risk posed by nuclear power but also metaphorically portrays life as stunted, even hopeless, after nuclear development. Often, the documented mutated plants are only subtly altered and need to be carefully studied by scientists to see signs of mutation; consequently, the imperceptibility of the deformed plant represents the insidious and invisible nature of the radiation threat: it is not always perceivable. Mutated nature as art is also recorded in the work of Tim Mousseau and Anders Moller who photograph abnormal Chernobyl and Fukushima insects, and the watercolors and illustrations of Cornelia Hesse-Honegger's mutated insects of Chernobyl.[35] Hesse-Honegger has also collected, cataloged, and drawn deformed bugs from the Windscale and Three Mile Island areas. Her famous American images include watercolors of a Harlequin bug, an Ambush bug, and Ladybird beetle, all malformed and taken from the site of Three Mile Island in 1991. Her other American nuclear insect art consists of deformed critters from near the Peach Bottom Atomic Power Station, near nuclear testing grounds, and from around Hanford, Washington (a former nuclear production complex).[36] The provocative images detail insects that visually appear distorted due to abnormal growths, stunted feelers, crippled limbs, discoloration, and asymmetrical features.

Such findings after nuclear disasters provoked a spike of environmental films in the 1990s and 2000s. Many cover the nuclear issue (such as *Pandora's Promise*, 2013), but others cover everything from air pollution in *Bhopal Express* (1999), waste management in *Polluting Paradise* (2012), water pollution in *Before Vanishing* (2005), and general environmental issues such as deforestation, global warming and species extinction in *11th Hour* (2007). Films such as *Renewal* (2008), an eight-piece documentary film, concentrate on environmentalism in the United States, which has become more prominent since the mid twentieth century. A contemporary documentary example of the uncomfortable convergence of the nuclear and environmentalism can be seen in *The Return of Navajo Boy* (2000), in which uranium mining is seen to not only impact the earth but the families gathered in those areas. This film examines uranium contamination on native American land and the damaging effects this had on Navajo families.

The Return of Navajo Boy responds partly to Kerr-McGee's propaganda film *A Navajo Journey* (1952), which boasts the optimism and wonder of the Nuclear Genie by tracing the exciting excavation of uranium. *A Navajo Journey* sees Native American families working in uranium mines to harvest fuel for nuclear plants. In the film, uranium is plundered using the resource of local labor by the now controversial and notorious Kerr-McGee Industries.[37] Kerr-McGee describes the Navajo people as skilled and spiritual and links the new practice of uranium mining to Navajo craft alongside wholesome practices of farming and jewelry making.[38] In fact, Kerr-McGee's promotional film dedicates sixteen minutes of its twenty minutes to outlining the Navajo way of life with special attention to the relationship between the people and the land. Kerr-McGee positions uranium mining as a modern way for the native community to engage with the earth.[39] The implication, in the final few minutes of *A Navajo Journey*, is that the nuclear industry has gifted the native people with modern machinery and modern enterprise which has enriched individuals and tribes as a whole. Yet, the respect the industry claims to have for the native people is unconvincing considering the mines are left spilling contamination into the local communities which poisons wells and hogans. *The Return of the Navajo Boy* seeks to draw attention to the environmental injustice faced by the Navajo nation. *The Return of the Navajo Boy* documents how years after Kerr-McGee's mining operation and the promises and optimism of the 1950s 'modern' uranium mining, the Environmental Protection Agency was forced to demolish hogans after discovering abnormal levels of radiation in materials used to construct homes near the mines. By the turn of the twenty-first century, the whole industry was under critique, from the beginnings in the uranium mines, to the meltdowns, and finally to the waste left behind at the end of the process.

WRITING MISTRUST: QUESTIONING AUTHORITIES, AND CYNICISM TOWARDS INDUSTRY

Environmental issues combined with media tales of conspiracy and criminality by the authorities, government, and industry created a strong vein of skepticism and cynicism in the 1980s towards nuclear power. The enduring skepticism of Karen Silkwood's accidental death is one example of how popular culture was shining an investigative light on woolly issues.

Renewed criticism also saw the republication of older anti-nuclear texts. The fact based graphic narrative *Nuclear Power for Beginners* by Stephen Croall and Kaianders Sempler was originally published in 1978 but was reprinted in the 1980s following Three Mile Island. The front cover instantly positions nuclear power as perilous through the misshapen skull on the cover and the cynical introductory illustration which portrays a captain parachuting away from a nose diving "Atomic Airways" plane. As the plane plummets, the 'nuclear captain' shouts "so long suckers".[41] The cartoon highlights, through a flight metaphor, that those onboard "Atomic Airways" are prisoners with no escape and are stuck on a flight deemed unsafe, untested, and uncertain. On the subject of Windscale, the crisis is

SILKWOOD AND THE "MALICIOUSNESS" OF INDUSTRY AS PRESENTED TO THE PUBLIC

The Silkwood trial that seemingly wrapped up in 1979 dragged on with a settlement that was agreed in 1986. Sara Nelson, working on the Silkwood case, boldly declared: "If people are injured by this industry's recklessness and maliciousness, and if they want to move against it through state law, they can."[40]

described as inevitable because nuclear disaster is "the nature of nuclear power".[42] However, the text outlines the 1970s as the decade in which America "woke up to the dangers of nuclear power".[43] Reflecting on the combination of human error and technical failure, Croall and Sempler argue that "human beings and nuclear power seems a lethal combination".[44] The text, presented almost as a textbook, combats educational texts published by the utilities and argues that "no nukes is good nukes" because it is better to be "active today than radioactive tomorrow".[45] The updated 1983 text includes information about the Three Mile Island accident. Three Mile Island became an example, the authors argue, of "a close shave" that could happen again—a sobering message when juxtaposed with a map of nuclear power reactors in America. Throughout the text nuclear power is presented as inherently unsafe, and anti-nuclear attitudes are promoted as a sign of intelligence. *Nuclear Power for Beginners* was formerly titled *The Anti-Nuclear Handbook*, the suggestion being that to learn about nuclear technology is to be anti-nuclear.

Even fictional texts, which portray families as brave in the face of nuclear crisis, carry derisive undertones against 'naive' nuclear-supporting citizens. Texts like *The Day After* (1982), *Testament* (1983), and *Threads* (1984) reflect on earlier sentiments of the 1940s, 1950s, and 1960s of nuclear deference by presenting innocent and gullible families optimistic or unwise about the disaster occurring around them. Naive optimism in such texts is not necessarily a commentary on the gullibility of people; rather, it is a reference to the suggested misdirection of the nuclear 'dream' and the selling of an ideal that was not always as peaceful as glorified. This is the case in the 1986 animated film *When the Wind Blows* (based on the graphic novel by Raymond Briggs), which follows the final days of an elderly couple who perish despite following the guidance in government pamphlets. The couple's belief that if radiation cannot be seen or felt it is not dangerous, results in the slow and painful death of both characters from radiation poisoning. They believe in the promises of the government literature until the very end. Even before Chernobyl, anxiety in the 1980s concentrated on the fear that radiation could decimate an area and that truth may not be conveyed by the authorities in the interest of maintaining community optimism.

Nuclear power was once conceived as part of 'Atoms for Peace' and characterized by the mystical and magical cartoon Nuclear Genie; however, in the 1980s it became symbolized in America by the Three Mile Island accident, with the inactive cooling towers acting as a new icon of the age. In turn, the name 'Three Mile Island' would become a phrase associated with the partial meltdown and nuclear accident. In Jonathan Sisson's 'The Crows of St Thomas' the poet mentions Three

Mile Island as a contemporary "*Titanic*".[46] In Yusef Komunyakaa's '1984' one stanza comments on suspected subterfuge surrounding Three Mile Island as well as mistrust towards authorities; the poem also suggests it will be a while until we know whether the radiation released was harmful. The poem mentions Three Mile Island's "dreamworld", perhaps suggesting that the dream of Three Mile Island is both the flawed promise of nuclear technology and the delusion surrounding claims that it is now 'safe'.[47] Molly McGrann dedicates an entire poem to the 1979 incident and entitles the piece 'Three Mile Island'; the final line of McGrann's poem comments on the feebleness of arguments surrounding nuclear technology which defend and excuse accidents.[48]

1980s fiction often took a venomous stance towards the industry, accusing those in charge of intentionally lying. Consequently, threat shifted from physical ramifications of disaster (explosions, radiation, fallout, contamination, looting), to the perils of living in a world, like Karen Silkwood, in which trusting industry and authorities is a risk. *The China Syndrome* made trust, or the lack thereof, one of the principle themes of the film even before the movie hit theatres. The trailer focuses not on a nuclear crisis but on mistrust by introducing the film as being about "people who lie, and people faced with the agony of telling the truth".[49] Buzzwords 'risk', 'lie', 'truth', and 'accident' act emotively alongside images of control rooms, sounds of alarms, and reoccurring blue speckled images reminiscent of atoms. The trailer ends with the ominous words: "Today only a handful of people know what it really means, and they're scared. Soon you'll know what it means."[50] The theme tune 'Somewhere in Between' by Stephen Bishop speaks of people caught 'in between' reality and distortion, which mirrors the film's central theme about media and industry subterfuge. The song, which also features a line about "drifters" who "lost their direction", seems to speak of the supposedly misguided individuals in the nuclear industry. Weeks later, the Three Mile Island accident occurred making these words oddly prophetic. Overall, the majority of the film involves scrutinizing how the industry responds to the accident—the mistruths fed to the media, the rushed and biased investigation, inadequate quality assurance, gross human error, shoddy procedures, cover-ups including attempted murder, and the ineffectual nature of the Nuclear Regulatory Commission (NRC).

Written after Three Mile Island, and inspired by it, the plot of Wentzell's *Faded Giant* centers around the idea that the locals have not been ill-advised, or misled, but point blank lied to intentionally. As mentioned earlier, *Faded Giant* portrays the Three Mile Island crisis as a full meltdown that was concealed from the community. Pelletier's realization of the severity of the accident, as well as discovering the cover-up, results in him being framed for murder. *Faded Giant* portrays the industry as determined to go to extreme lengths to protect its reputation and profits. Pelletier's realization that "I don't think anyone wants to know the facts" summarizes conspiracy suspicions surrounding both the Silkwood case and nuclear accidents.[51] However, the spiral of decay in the industry extends beyond lies as Wentzell's novel is littered with references to immoral and criminal behavior by all plant employees. Workers are dangerously incompetent (the novel opens with Three Mile Island technicians accidentally tracking contaminated water through a publically accessed motel) and of questionable morality (many are reported as alcoholics, drug users, and many visit prostitutes). The industry is not only aware of these behaviors but seems to be encouraging them by seducing the workers into silence over accidents with happy hours and bonuses (to be spent on prostitutes and marijuana).

Ideas of corruption in the industry provoked many texts that directly commented on nuclear power and the nuclear industry to describe the two as 'evil'. The 'evil' in industry is perhaps best seen in fiction through the portrayal of EduComics' Greedy Killerwatt—a lightbulb based comic book character. This demonic lightbulb is a satirical and dystopian play on the Reddy Kilowatt brand character used in electricity utility advertisements.[52] EduComics' *All-Atomic Comics* (1976) attempts to educate the public on dangerous nuclear power—the colorful front cover of issue 1 depicts Greedy Killerwatt ominously answering a young boy's question "is nuclear power the answer?" with the sinister retort that he would "bet your *life* on it!" After Three Mile Island, Marvel borrowed Greedy Killerwatt from *All-Atomic Comics.*

Marvel inserted Greedy Killerwatt into the Three Mile Island inspired 1980 Christmas issue of *Howard the Duck*. Entitled 'A Christmas for Carol', edition creators Bill Mantlo, Gene Colan, and Dave Simons combine residual anxieties over the Three Mile Island accident a year before with a traditional Christmas story about a child trying to rediscover the magic of Christmas. The climax of the story focuses on the fictitious North Pole Nuclear Facility. The story goes that Santa Claus has chosen to install a nuclear power facility at the North Pole in an effort to produce modern toys quickly and cheaply. The plant and toy factory is then taken over by the evil Greedy Killerwatt. Santa's sacrifice of quality and safety in his workshop for cheaper and quicker production is an obvious allegory for the anti-nuclear claims that 'too cheap to meter' nuclear power is a dangerous con that did not live up to expectations and ultimately put children and families at risk. Despite the tyranny of Greedy Killerwatt, it is clear that Santa is to blame for the dangerous plant because, as he acknowledges, he knew the risks but proceeded anyway. He also believed that if an accident did occur no one would be hurt due to the plant's location at the North Pole. However, drawing satirically on *The China Syndrome*, Santa did not take into account "The Antarctica Syndrome" (which would produce the same earth melting effect). Howard, the hero of the piece, debates saving himself and leaving the human race to face certain death based on the rationale that their naivety caused this peril and thus it is their fault to some extent.

Although Santa is to blame for the presence of the plant, the villain of the piece is Greedy Killerwatt. As we discover in the comic, Greedy Killerwatt is a mutated evil lightbulb who was once a human Three Mile Island employee paid to be a 'jumper' (employed to enter hot areas to do maintenance). He was exposed to high doses of radiation at Three Mile Island and after becoming sick with radiation poisoning, his body mutated into an incandescent mass in the form of a humanoid lightbulb. Rather than an anomaly, Killerwatt is joined by gruesome mutants that were once aquatic. The contagion of radiation, especially being linked to water, speaks to dominant fears of the time about the contamination of areas like the Susquehanna. Linking nuclear power to nuclear weaponry, Killerwatt threatens to use the plant as a weapon in a lyrical rant:

> IF YOU WANT NUKES I GOT A LOT,
> THE STUFF I PUSH IS REALLY HOT,
> THE POWER OF TOMORROW!
>
> I'M THE GUY WHO RAISES CANE,
> POLLUTES THE FISH, POISONS THE RAIN,
> BUT IF YOU WANT WATTS YOU CAN'T COMPLAIN,
> MY LAUGHTER IS YOUR SORROW.[53]

Here Killerwatt draws attention to the influence the industry has over those who demand constant wattage and how this can be supplied to their detriment. Eventually, the crisis is averted and Killerwatt is vanquished. In the final stages of the narration, Santa turns to solar power.

In this example, not only is the industry depicted as inherently evil, but those who support it are depicted as obedient fools who value power over safety. Publications like *Corporate Crime Comics* and *Faded Giant* portray mistrust over the 'shadowy' behavior of the industry. Texts like *Nuclear Power for Beginners* speak of the natural danger of the industry manned by 'captains' calling citizens "suckers", while films like *The Day After*, *Testament*, and *Threads* suggest an innocent naivety of those facing a nuclear threat. While texts like *When the Wind Blows* question the naivety and optimism of government assurances towards the dangers posed by nuclear technology, 'A Christmas for Carol' is more venomous in its approach by casting the industry as evil. These texts, post Three Mile Island, question the extent to which authorities can be trusted and to what extent misinformation has problematized citizen knowledge. Such complaints in fiction were not new but the 1979 incident, in an original way, questioned the dominance of an industry that had been linked to subterfuge (not helped by additional conspiracies like Silkwood and Kerr-McGee's Navajo betrayal).

FICTION AS MICROCOSM: *THE SIMPSONS* AND TECHNOLOGY, HUMAN ERROR, AND INDUSTRY CYNICISM

One important and enduring cultural example which features the hauntology of Three Mile Island is demonstrated in American prime-time family favorite *The Simpsons* (1989–present). Aimed at both children and adults, *The Simpsons*, created by alternative comic artist Matt Groening, has been no stranger to tackling socially relevant issues and satirizing suburban America. The issue of nuclear power is both an ongoing commentary and joke in this animated show. Despite the show's focus on an all-American family in a small suburban town, the Springfield power plant (a pressurized water reactor with two cooling towers) plays a major role as demonstrated by the title sequence. The show opens with a shot of the power plant on a hill overlooking Springfield and positions industry as a significant component of American life; domestic scenes at the school and supermarket are interrupted with clips of protagonist Homer Simpson working with plutonium and accidentally transporting it home. In many respects, *The Simpsons* acts as a microcosm for all the anxieties raised by the 1979 accident: technological failure at the nuclear plant, human error by plant workers, community endangerment, the effects of danger (such as radiation) on the family unit, environmental issues, cynicism towards the industry, and large-scale crisis.

The plant is owned and managed by town tyrant Mr. C. Montgomery Burns who is unconcerned about the safety of the people and his employees. Simpson is an inept, negligent, and dangerous employee who lacks sufficient training and skill to be a plant safety officer. Many episodes show the plant as dangerous, poorly run, and operated by inattentive employees. In fact, the plant is so dangerous that Burns has an escape pod. Workers are also portrayed (like in *Faded Giant*) as lazy alcoholics; many sleep at their posts which metaphorically says something about the 'dream' of nuclear power and the suggested obliviousness to its 'nightmarish' reality. In 'Homer Goes to College' all workers are asleep when a core meltdown is triggered (Simpson falls on the 'Plant Destruct' button while asleep). Despite alarms sounding, none of the workers are alerted to the problem. Such obliviousness and detachment are emblematic of inherent nonchalance that typifies most Springfield workers. In the mobile application game *The Simpsons: Tapped Out*, Simpson is distracted at work again (this time playing on an iPad) and his inattention leads to a meltdown resulting in a massive explosion and giant mushroom cloud.

Yet, the ineptitude of the workers is not solely down to indifference, many employees, like Simpson, are not adequately trained. Simpson's incompetence is worsened by his inability to understand and successfully co-work with plant technology. In the first episode aired in December 1989, Simpson is seen inspecting plant readings, signing off on each one despite panels recording 'danger' and flashing warning lights. In 'King Sized Homer', an exchange between Simpson and a computer highlights the superiority of complex technology over undertrained operators:

> *Computer:* Vent radioactive gas?
> *Simpson:* No
> *Computer:* Venting prevents explosion.
> *Simpson:* Ooh, this is hard . . . Okay, then. Yes.[54]

Believing that the computer can handle all his responsibilities, Simpson constructs a crude automated system which allows him to select 'Yes' to all the computer's questions. This allows him to leave his station negligently for extended periods. However, his system fails and upon his return he finds that the plant is in a critical state. Simpson narrowly avoids disaster by venting radioactive gas manually. His stupidity remains undetected and he is rewarded with a medal.

Due to the sheer number of accidents and disasters, the plant has seen many inspections, all of which are exposed to be inadequate or corrupt. In 'Homer Goes to College', Burns attempts to manipulate, trick, and bribe the NRC and tries to hide a multitude of problems with the plant and its workers. In 'Two Cars in Every Garage and Three Eyes in Every Fish', Burns attempts to bribe government inspectors when faults are found at the plant. The list of faults includes a crack in the cooling tower sealed with gum, a plutonium rod used as a paperweight, an unmanned control room, nuclear waste left in barrels in the staff room, and leaking pipes—amounting to 342 violations of code.[55] Inspectors also visit after a mutated three-eyed fish is found in the water around the plant. The inspectors tell Burns that they have not seen a plant so badly run and so dangerous for twenty years, a nod to Three Mile Island in the 1970s. Burns' expertise in misdirecting authorities, inspectors, and the media creates considerable ambivalence in the local community; the plant, despite many obvious failings and very public disasters, is usually ignored by townsfolk due to the placating promises offered by the plant owner. In Season 2, Burns speaks to television reporter Kent Brockman after a meltdown; during the exchange, he demonstrates control over the media:

> *Mr. Burns:* Right now, skilled technicians are calmly correcting a minor, piffling malfunction but I can assure you and the public that there's absolutely no danger whatsoever. Things couldn't be more shipshape.

Burns delivers this statement while donning a radiation suit as plant workers panic in the background. He further lessens the gravity of the situation when the interviewer calls the crisis a "meltdown" by replying "That's one of those annoying buzz words. We prefer to call it an 'unrequested fission surplus' ".[56] Typical to form, the crisis is averted through sheer luck; however, Burns dismisses the whole disaster as a false alarm to which Brockman replies: "Your point about nuclear hysteria is well taken. This reporter promises to be less vigilant in the future."[57] This exchange corresponds to claims during the Three Mile Island accident that many official reports were intentionally confusing or manipulative. The feminist publication *Off Our Backs*, in May 1979, described deliberate "patronizing platitudes, cover-ups and confusion" surrounding Three Mile Island.[58] Also, there was some suggestion that terminology was used intentionally to downplay the severity of the crisis. *Time* notes that while the industry referred to Three Mile Island as an "event", the reality was that America was faced with its worst nuclear power accident in nuclear history.[59] Burn's exchange with Brockman is also reminiscent of how the NRC uses language; when approached by Robert Del Tredici about "accidental releases" at Three Mile Island, John Collins of the NRC replied, "I don't particularly care for the phrase 'accidental releases.' I would characterize them as 'events' ."[60]

As nuclear power is not comprehensible by many who work at the plant, the townspeople, largely, have no understanding of the technology and thus do not have the tools to engage in debate or challenge authority. Thus, the corruption and danger posed by the nuclear industry are frequently ignored. Another reason why the plant is largely unopposed is due to the reliance the town has on the energy produced. In 'Last Exit to Springfield' Burns makes it clear that any challenge to him or his business will result in the severance of all electricity.[61]

To engender local loyalty, the positive aspects of the nuclear plant are solidified in the residents from a young age. The children are conditioned through comic books and film to find mystery and magic in the atom through their heroes Radioactive Man and Fallout Boy—just like the audience's superheroes (Captain Atom, Spiderman, Dr. Manhattan). School trips to the plant have children watching a parody of an Atoms for Peace video akin to *Our Friend the Atom* called 'Nuclear Energy Our Misunderstood Friend' narrated by Smilin' Joe Fission (a nod to Reddy Kilowatt). Nuclear power is also positively positioned through the dominance of popular culture in the lives of the townspeople. As Mick Broderick assesses, "atomic culture is synonymous with popular culture" on the show as characters often make sense of atomic development and threat through how it is positioned in mass culture.[62] An example Broderick provides is how Bart imagines Marie and Pierre Curie's radium discovery through picturing scenes from Godzilla.[63] Here, Bart finds pleasure and detachment from the tragic nature of the Curies' demise through sensationalism (which was the aim of some nuclear Hollywood films of the 1950s). Simpson, like all Springfield citizens, is heavily influenced by television (which we know evolved alongside nuclear technology). He passively resides on his couch waiting to be told what to think and feel which illustrates the permeation of popular culture. Such control is exemplified in Season 6, Episode 10 when Grampa Simpson finds an old TV set called the Radiation King (a nod to the fact early sets gave off radiation) which has scorched a shadow of a television viewer on the wall. The obvious connotations of Hiroshima not only extend to the iconic 'Nuclear Shadows' permanently stained on Hiroshima's steps, floors, and walls but suggest that nuclear technology has left a dark figurative shadow over everyone in the nuclear age. In this episode, Simpson is depicted sitting in the same position as the 'radiation shadow', cementing the metaphor that television (and by extension popular culture) leaves a very real impact on each viewer. Thus, through popular culture, especially visual culture, nuclear power in *The Simpsons* (and even nuclear weapons in several episodes) saturates life in the small community.

WRITING APOCALYPSE: WAR MAKES WAY FOR MELTDOWN

Lifton notes that with "the two American atomic bombings in 1945 there came into the world a special image: that of exterminating ourselves as a species with our own technology".[64] The focus here is on nuclear power and, obviously, nuclear power does not instantly conjure the same apocalyptic imagery or potential as weaponry. However, radiation acts as a link between nuclear war and nuclear energy. The idea of a meltdown, as seen with accidents like Chernobyl, can help people visualize a localized, even personal 'apocalypse' limited to a region rather than a continent. Award winning documentary *Dark Circle* (1982) suggests that war-type scenarios are not the only way nuclear technology can jeopardize America and the globe. Focused on the Rocky Flats Plant and filmed in America and Japan, *Dark Circle* begins its anti-nuclear discussion with the ominous quote "if you are working with a radioactive substance don't let it escape. If you let it escape into the atmosphere it might come back to you."[65] Escape is the theme of *Dark Circle* in many respects; the human race may have escaped a nuclear World War III but has not escaped the nuclear threat in the form of nuclear plants dotted throughout the country, it suggests. *Dark Circle* partly examines the government's Idaho 1955, Nevada 1965, and New Mexico 1965 projects in which reactors were intentionally destroyed to observe how plant disaster would affect the locale. As mushroom clouds in startling gold and red billows burst from the site, the narrator reminds the audience "this is not an atomic bomb; this is the actual explosion of an unshielded nuclear reactor".[66] The 'dark circle' is a circle that links nuclear weapons to nuclear power. Reflecting on the Rocky Flats Plant for nuclear weapons production, activist Pam Solo associates the weapon with potential contamination and destruction that can come with power plants; weapon and plant then fuse together to represent "the end of some kind of continuity".[67] Weapons for Solo risk human annihilation, but the plant on a local scale risks terminal cancer.[68] Both scenarios are apocalyptic for anyone experiencing the crisis.

An interesting convergence of nuclear weapon with nuclear power is evidenced in an August 1980 edition of *Peace Newsletter* (the Central New York's Antiwar/Social Justice Paper) in which Hiroshima is housed alongside Three Mile Island. The newsletter advertised Hiroshima-Nagasaki Week to mark the thirty-fifth anniversary of the atomic bomb drop on Japan in 1945. The full-page advertisement features a large stop sign covering the silhouette of a nuclear missile. The downwards facing missile points to the headline "we are the guinea pigs: Three Mile Island and the Catastrophe of Nuclear Power [sic]".[69] This reference is to the film of the same name, also being shown on the anniversary of Hiroshima. The newsletter links anti-war and anti-nuclear power as part of a social justice movement. Both power and weapon are united as a collective nuclear threat. This is further solidified on another page where a skeleton gleefully dances between a nuclear missile and a nuclear power plant, captioned with "Zero Nuclear Weapons. Stop Nuclear Power".[70] In the accompanying article, Geoffrey Navias speaks of the importance of uniting all anti-nuclear movements and considering them to be one issue because they are united in sharing the same cause and solution.[71] Three Mile Island became one important symbol of the fearsome potential of the nuclear, as illustrated in this publication. Although crisis was averted, an example of what could go wrong had been highlighted graphically in 1945 and thus Three Mile Island became interconnected with nuclear danger in a more general sense. Publications like *Peace Newsletter* suggest that the barren, radiation-drenched landscape of Hiroshima could have been a reality for the poor folks living on the doorstep of a nuclear reactor in Harrisburg.

As we saw in Chapter 2, the American government tried to reposition nuclear technology into a benevolent force through nuclear power. But, despite the efforts of Disney and the government, there remained a dark legacy of the nuclear cloud even when discussing 'safe' and 'domesticated' nuclear power technologies. Cynical awareness arose alongside the main thrust of the anti-nuclear movement and peaked after nuclear power accidents. It is the fear of the radiation trace the cloud leaves behind that truly unites these very different events. Although not a mushroom cloud, the power industry has been associated with numerous artificial clouds of contamination. Windscale, England, not only released clouds of smoke during the 1957 accident, but it released clouds of radioisotopes, concealed from the people but manifest when contaminated milk had to be destroyed. Years later, when discussing the 1979 nuclear power plant accident at Three Mile Island in Pennsylvania, America, Lonna Malmsheimer linked the disaster to the "first icon of the nuclear age, the mushroom cloud".[72] In Halevy's account of Three Mile Island in her memoir *Yes, I Glow in the Dark!*, she remarks on fearsome clouds when she recalls her terror at the fog outside her home: "There was a smog-like haze in the air, thick and humid. Caused by radiation? Carrying fall-out? How many roentgens in a millirem?"[73] The toxic cloud that was released during the Chernobyl 1986 disaster was detected across Europe on radiation detectors. An article in *New Scientist* (1987) aptly describes the importance of traces left behind in this cloud story: "The physical cloud, although not its radioactive burden, may have dissipated, only time will lift the cloud of fear that hangs over nuclear power."[74]

Before Three Mile Island, the idea of an apocalyptic nuclear scenario was almost completely limited to a war-type event. After 1979, more creative products were issued focusing on the nuclear power industry contributing to devastation. Partly, the shift from a nuclear war to a nuclear power accident as being apocalyptic was assisted by three main things: first, the shock of the Three Mile Island accident on American soil; second, the close association between nuclear weapons and nuclear power since the 1940s; finally, the close of the Cold War and the decline (although not end of) nuclear activism, protest, and criticism over nuclear weaponry/war, thus leading to a renewed focus on power as the primary concern. *The China Syndrome* perfectly summarizes the second point. In the film, Richard Adams comments that the uranium used to power nuclear plants is also used to make nuclear bombs. As Kimberly learns about nuclear reaction from an engineer at the plant, industry terms like 'chain reaction' (along with Adams' comment about bombs) may remind the viewer of the chain reaction experienced during nuclear detonation—the reaction that leads to nuclear explosions and mushroom clouds. Further, filmic control rooms in both nuclear plants and military establishments are uncannily similar in design. Both types of control room are equally perplexing to non-expert audiences: circular rows of lit computer screens, unintelligible readings, flashing lights, consoles, and buttons. Three Mile Island, as the worst accident in industrial history in America, made the nuclear power crisis an American concern, whereas before it 'belonged' to the Soviet Union (Urals) and the United Kingdom (Windscale). However, the fact that crisis was ultimately averted painted the event as aborted in one sense, and in another sense the event became fodder for 'What If' scenarios in fiction. Paradoxically, Three Mile Island in dystopian apocalypse stories acts as a 'pre-event' that imagines a disaster like Chernobyl. This is one reason Chernobyl often overshadows Three Mile Island as a narrative tool—Three Mile Island imagined disaster, but Chernobyl actually delivered it.

Nevertheless, Three Mile Island changed the way America viewed nuclear power. Donald Neff, from the New York Bureau, reflected on his childhood holidays to Goldsboro, commenting that the Island once held mystery but after the accident was transformed into something "ominous and frightening".[75] This transformation of American towns and American landscape from mysterious to uncanny sows the seeds of apocalyptic potential in fiction. What does it mean to live in a world of the "ominous and frightening"? Although *Life* magazine explained to readers that the crisis was over, and fear was abated, the crisis worsened and bloomed in fiction as a multitude of scenarios played out. From paranoid engineers fleeing the deadly bite of radiation in *Faded Giant*, to the tyrannical nuclear industry represented through the villainous Greedy Killerwatt, to the story of a local family in *My Father, the Old Horse*, Three Mile Island has featured explicitly in many stories.

Michael Swanwick's novel, *In The Drift*, features the most apocalyptic story involving the Three Mile Island crisis to date. Written in 1981, long before Chernobyl, the plot imagines that the Three Mile Island accident had ended in a complete meltdown that scorched the sky and transformed the landscape into a barren wasteland saturated with radiation. This area, once home to Harrisburg, is now called 'The Drift' and is toxic to anyone who dares venture within and is the home of mutants and vampires. Healthy refugees fled the area in 1979 and headed for Philadelphia, having been told that the area was safe. However, Swanwick reflects on the real issues of distrust experienced during the actual accident by having his fictitious government hide from the people the fact that Philadelphia is also not safe. The contamination has spread and, through the outlawing of high-technology, the people are unable to detect radiation in the safe zone. Those who stumble upon the truth are murdered by a specially created task force.

Unlike *Beaver Pig*, the mutations experienced in *In The Drift* are horrifying with many scenes featuring grievously mutated children, and adults suffering from miscarriages and crippling radiation sickness. Genetics Laws in place outside The Drift provoke (and authorize) the hunting and killing of anyone with genetic diseases; the hunt is even part of a festival. Vigilante justice is also witnessed as those outside The Drift are prepared to neutralize any threat coming from the contaminated land—even if this involves the killing of citizens trying to evacuate. Those who are genetically damaged or politically dangerous are transported to The Drift and exiled. Unlike *Faded Giant*, which also involves a full meltdown, Swanwick's novel is a chilling apocalyptic, dystopian thriller. In such a dystopian world, religious prayers and rituals speak of radiation and genetic disease which reflects the shift from religious to secular apocalypse experienced in the twentieth century as noted earlier. Such a transformation is highlighted by protagonist Keith when he discovers what looks like a modern shrine in The Drift: "Where a crucifix should have been, there was a bright, crudely painted radiation logo."[76] In a ritual inspired by Native American and Catholic ceremonies, objects like lead suits become sacred artefacts and the 'nuclear colors' of red and yellow (the color of warning signs) take on special protective significance. The terms 'evil' and 'hell' are now linked to the reactor that resides in the center of the meltdown zone and this even takes on a degree of sentience as it is described as both an animal and an all-seeing eye. On the back cover of the 1985 edition, American author Gardner Dozois speaks of the novel as potentially prophetic by claiming that *In The Drift* shows a conceivable future reality. It is doubtful that Dozois believes that vampires and mutants could roam the Earth after a nuclear accident; instead, it is likely that the reviewer read about the contamination, hysteria, and grievous environmental toll and conceived of this as possible. Chernobyl, a year after the publication of this edition, would prove that point.

CONCLUSION: THE CONVERGENCE OF FICTION AND REALITY

Throughout the chapter, we have seen examples of how the 1979 Three Mile Island event has implicitly or explicitly informed creative work after

Three Mile Island in the Background of Twenty-First Century Blockbusters

Even when the plant does not take a leading role in a fictional account, the presence of the infamous nuclear power station lingers in the background of many texts, a looming presence much like the haunting imagery of the ominous mushroom cloud. Even in the twenty-first century, Three Mile Island is used as a cultural symbol in the background of many films such as apocalyptically themed American blockbuster hits *X-Men Origins: Wolverine* (2009) and *Super 8* (2011).

Case Studies: *Super 8* and *X-Men*

J. J. Abrams' award winning hit *Super 8* (2011) features a news report warning about a potential meltdown at Three Mile Island which sets the film precisely in 1979. The film partly features shady actions by the American military who seek to cover up the particulars of a suspicious rail accident in Ohio. In several scenes, the military is seen checking the crash site and town with Geiger counters. Town hall meetings are held with the locals questioning the authorities about strange goings-on in the community and an invisible force that is seemingly targeting the area. The eventual power outages and mandatory evacuation of the town coincide with growing distrust of the military presence and the belief that they have accidentally unleashed a dangerous monster that cannot be controlled and is able to attack through invisible means. These themes will surely remind us of the anxieties covered in the last chapter about the invisible threat of radiation and a mistrust of authorities. Furthermore, the creature the military hunt is subterranean and burrows holes deep into the earth; these holes do not quite reach all the way to China—but the *sinking* effect is notable.

Another example is the high grossing *X-Men* spinoff, *X-Men Origins: Wolverine* (2009). In the film's climactic moment, mutant hero James 'Logan' Howlett (Wolverine) battles mutant super villain Weapon XI (Deadpool) on top of one of Three Mile Island's cooling towers. During the fight, one of the cooling towers is destroyed, leaving behind substantial ruination. The setting of Three Mile Island and its complicated history enhances two dominant themes in the film: cynicism towards the actions of the government/authorities, and the issue of genetic mutation. Radiation has been known to cause DNA damage, and the entire *X-Men* franchise is based on genetic mutation. Cynicism towards the government surfaces once more as the secret laboratory hidden at Three Mile Island is run by the corrupt and reckless Major William Stryker, leader of the government's Weapon X project (involving turning mutants into weapons).

Figure 4.1 Three Mile Island in *X-Men Origins: Wolverine* (2009). Three Mile Island is under attack as Deadpool demolishes one of the cooling towers

X-Men Origins: Wolverine, dir. by Gavin Hood (20th Century Fox, 2009)

the crisis. *The China Syndrome* has acted as a principle text through which to explore key themes, even though this film came before the real event. The importance of this text, in particular, is something philosopher Jean Baudrillard has reflected on. In *Simulacra and Simulation*, Baudrillard wrote an essay on the convergence of the televisual and the Three Mile Island incident called '*The China Syndrome*'. Baudrillard's argument is that the fictional event has actually overtaken the reality of the crisis itself; he argues that "*The China Syndrome* is a great example of the supremacy of the televised event over the nuclear event which, itself, remains improbable and in some sense imaginary".[77] Baudrillard not only suggests that the film ended up dominating our understanding of Three Mile Island, but that the accident itself became illusory due to it also being part of a televised spectacle. The presentation of Three Mile Island as a 'story' in numerous publications, using (as mentioned earlier) creative phrasing akin to novel writing, alongside a wealth of challenged reports, rumors, and untruths, problematizes the authenticity of the event and the relationship between the citizen (now viewer/audience) and the crisis. Here, the Three Mile Island incident is more complicated than perhaps initially understood and this is to do with its connection to fiction. *The China Syndrome* prepared locals, and wider America, for Three Mile Island (as Baudrillard terms it, the film was an *induction* for the people). However, Baudrillard also suggests that the dominance of the fiction (including media focus and misinformation) considerably problematizes any truth about the accident. Can we ever know or understand Three Mile Island as anything other than a story? There are no ruins. There was no explosion. There was no visual cloud or fiery spectacle.

One of the reasons Three Mile Island was so influential was that we were seemingly waiting for it; as Baudrillard stresses, we anticipate disaster: "The equilibrium of terror rests on the eternal deferral of the atomic clash."[78] Yet at the same time, such a crisis was never *supposed* to occur, even if it can be imagined, and especially not in America, a country positioned as the father of atomic power. "*Nuclear catastrophe does not occur, is not meant to happen*", says Baudrillard.[79] Partly, the idea that a nuclear catastrophe will not and should not occur is because the nuclear weapon has been repositioned as a deterrent and nuclear power has been positioned as a safe and innocent alternative use of nuclear technology. Thus, when disaster does strike it is all that more jarring and has resonance for decades to come.

Three Mile Island was famous for being the worst commercial industrial accident in American history; but it is also famous because a full meltdown did not happen. Likewise, in *The China Syndrome*, a nuclear disaster does not occur. Baudrillard reflects on this when he speaks of the human drama taking precedence in the film, and the eventual showdown between Jack Godell and plant officials taking the place of a meltdown or explosion. Both *The China Syndrome* and Three Mile Island created anticipation for an apocalyptic disaster that did not occur. Why did *The China Syndrome* balk at exploring apocalypse? Ultimately, *The China Syndrome* wanted to explore human issues and draw attention to a range of concerns surrounding the nuclear power industry, rather than descend into the traditional nuclear horrors that dominated the 1950s. Also, the film came before the horrific spectacle of Chernobyl and with Windscale as the only example of grievous plant failure, so the creative leap from contaminated milk (Windscale) to the ruination of a large area with a substantial death toll (Chernobyl) was not secure. Three Mile Island helped bridge this conceptual gap through 'What If' scenarios. Chernobyl and Fukushima would actualize the apocalyptic crisis that Three Mile Island imagined but did not complete.

NOTES

1 Reported in Richard Roberts, 'Panel Probes Health Aspects of TMI Accident', *The Patriot*, May 23, 1979, p. 31.
2 David Marc, *Comic Visions: Television Comedy and American Culture*, 2nd edn (Massachusetts: Blackwell, 1997), p. 148.
3 Joyce Nelson, *The Perfect Machine* (Ontario: Between the Lines, 1987), p. 26.
4 Nelson, *The Perfect Machine*, p. 35.
5 Nelson, *The Perfect Machine*, p. 43.

6 Jacques Derrida is accredited with the origin of the term and theory. In *Spectres of Marx: The State of Debt, the Work of Mourning and the New International*, Derrida spoke of the haunting presence of Marx on Western thought. Here, I am using the term broadly and loosely to articulate a sense of the nuclear specter.
7 Bob Dvorchak and Harry F. Rosenthal, 'Seven Days at Three Mile Island Nuclear Power Plant', *Daily News*, April 8, 1979, p. 6A.
8 *Life*, 2.5 (May 1979), p. cover.
9 'A Nuclear Nightmare', *Time*, 113.15 (April 9, 1979), p. 8.
10 'A Nuclear Nightmare', p. 11
11 Lonna M. Malmsheimer, 'Three Mile Island: Fact, Frame, and Fiction', *American Quarterly*, 38.1 (Spring, 1986), 35–52 (p. 38).
12 *Inviting Disaster: Three Mile Island*, dir. by David DeVries (History Channel, 2003). The full quote offered is "the low ebb of the soul; late at night, often the time when accidents happen" paraphrased in the documentary from *Something Wicked This Way Comes* by Ray Bradbury.
13 Janet Kafka, 'Why Science Fiction?', *The English Journal*, 64.5 (May, 1975), 46–53 (p. 46).
14 Joanna Russ, 'The Image of Women in Science Fiction', *Vertex* 32.1 (1974), 53–57.
15 Lou Cannon, 'Actor, Governor, President, Icon', *The Washington Post*, 6 June 2004, p. A01.
16 *The China Syndrome*, dir. by James Bridges (Columbia Pictures, 1979).
17 *The China Syndrome Trailer*, dir. by James Bridges (Columbia Pictures, 1979).
18 *Panic in Year Zero!* dir. by Ray Milland (American International Pictures, 1962).
19 Libbe Halevy, *Yes, I Glow in the Dark! One Mile from Three Mile Island to Fukushima and Beyond* (n.p.: Heartistry Communications, 2014), Kindle ebook, LOC 335.
20 Benjamin C. Garrett and John Hart, *Historical Dictionary of Nuclear, Biological and Chemical Warfare* (Plymouth: Scarecrow Press, 2007), p. 73.
21 Timothy Wentzell, *Faded Giant* (Bloomington: AuthorHouse, 2009), p. 10.
22 Robert J. Lifton and Richard Falk, *Indefensible Weapons: The Political and Psychological Case Against Nuclearism* (New York: Basic Books, 1982), p. 13.
23 Films such as *Testament* (1983), *The Day After* (1983), *Def-Con 4* (1985), *Special Bulletin* (1983), and 'Shelter Skelter' from *The Twilight Zone* (1987); and literature including Luke Rhinehart's *Long Voyage Back* (1983), Whitley Strieber and James Kunetka's *Warday* (1984), Robert R. McCammon's *Swan Song* (1987), and William Brinkley's *The Last Ship* (1988).
24 *Chernobyl Heart*, dir. by Maryann DeLeo (HBO, 2003).
25 The film was shown at Elks Theatre in Middletown. Noted in 'Elks Theatre Will Mark TMI Anniversary with "My Father, The Old Horse" and "The China Syndrome"', *Pennlive* (May 5, 2014), available online at www.pennlive.com/entertainment/index.ssf/2014/05/elks_theatre_will_mark_tmi_ann.html (accessed August 25, 2016).
26 Martin Heidegger, *The Question Concerning Technology and Other Essays*, trans. by William Lovitt (London: Harper and Row, 1977), p. 23.
27 Liam Sprod, *Nuclear Futurism* (Hants: Zero Books, 2012), p. 80
28 Adam Scott Clark, *Beaver Pig* (Lancaster: unpublished manuscript, 2013), p. 2.
29 Clark, *Beaver Pig*, p. 3.

30 Clark, *Beaver Pig*, p.10.
31 Adam Scott Clark on *Beaver Pig* (email to Grace Halden, February 15, 2016, 21:12 p.m.).
32 Clark, *Beaver Pig*, pp. 259–260.
33 Clark, *Beaver Pig*, p. 295.
34 Lauren Redniss, *Radioactive: Marie & Pierre Curie. A Tale of Love & Fallout* (New York: HarperCollins, 2011), p. 103.
35 See Helen Thompson, 'Chernobyl's Bugs: The Art and Science of Life After Nuclear Fallout', *Smithsonian Magazine* (April 26, 2014), available online at www.smithsonianmag.com/arts-culture/chernobyls-bugs-art-and-science-life-after-nuclear-fallout-180951231/?no-ist (accessed May 18, 2016).
36 See Cornelia Hesse-Honegger, 'Field studies in radiation-contaminated areas in the U.S.', (n.d.), available online at www.wissenskunst.ch/uk/completed-studies/united-states/1/. (accessed May 18, 2016).
37 *A Navajo Journey*, dir. by C. J. Colby (Kerr-McGee, 1952).
38 *A Navajo Journey*.
39 *A Navajo Journey*.
40 'Silkwood Case Laid to Rest', *Science News*, 130.9 (August 30, 1986), p. 134.
41 Stephen Croall and Kaianders Sempler, *Nuclear Power for Beginners* (London: Beginners Books, 1978), p. cover.
42 Croall and Sempler, *Nuclear Power for Beginners*, p. 35.
43 Croall and Sempler, *Nuclear Power for Beginners*, p. 43.
44 Croall and Sempler, *Nuclear Power for Beginners*, p. 36.
45 Croall and Sempler, *Nuclear Power for Beginners*, pp. 120, 154.
46 Jonathan Sisson, 'The Crows of St Thomas', *Poetry*, 139.3 (December 1981), 139–141 (p. 141).
47 Yusef Komunyakaa, '1984', *Callaloo*, 20 (Winter, 1984), 114–118 (p. 115).
48 Molly McGrann, 'Three Mile Island', *Columbia: A Journal of Literature and Art*, 30 (Fall 1998), p. 198.
49 *The China Syndrome*.
50 *The China Syndrome* Trailer.
51 Wentzell, *Faded Giant*, p. 40.
52 Leonard Rifas, 'Cartooning and Nuclear Power: From Industry Advertising to Activist Uprising and Beyond', *PS: Political Science and Politics*, 40.2 (April 2007), 255–260 (p. 256).
53 Bill Mantlo, Gene Colan, and Dave Simons, 'A Christmas for Carol', *Howard the Duck*, 1.3 (New York: February 1980, p. 45.
54 'King Size Homer', *The Simpsons*, Fox, November 5, 1995.
55 'Two Cars in Every Garage and Three Eyes in Every Fish', *The Simpsons*, Fox, November 1, 1990.
56 'Homer Defined', *The Simpsons*, Fox, October 17, 1991.
57 'Homer Defined', *The Simpsons*.
58 Margie Crow, 'Nukes vs. Anti-Nukes: Malignant Monster Meets Critical Mass Movement', *Off Our Backs*, 9.5 (May 1979), pp. 2–6.
59 'A Nuclear Nightmare', p. 8.
60 Robert Del Tredici, *The People of Three Mile Island* (San Francisco: Sierra Club Books, 1980), p. 116. Collins speaks of expected releases and thus does not like

the term 'accidental releases' for they are expected and normal 'events'. Still, this exchange highlights how carefully language is selected.

61 'Last Exit to Springfield', *The Simpsons*, Fox, March 11, 1993.
62 Mick Broderick, 'Releasing the Hounds: The Simpsons as Anti-Nuclear Satire', in *Leaving Springfield*, ed. by John Alberti (Detroit: Wayne State University Press, 2004), pp. 244–272 (p. 246).
63 Mick Broderick, 'Releasing the Hounds: The Simpsons as Anti-Nuclear Satire', in *Leaving Springfield*, ed. by Alberti, pp. 244–272 (p. 246).
64 Lifton and Falk, *Indefensible Weapons*, p. 57.
65 *Dark Circle*, dir. by Judy Irving (Independent Documentary Group, 1982).
66 *Dark Circle*.
67 *Dark Circle*.
68 Another important example is *Deadly Deception: General Electric, Nuclear Weapons and Our Environment* (1992), which examines the environmental and human health dangers posed by nuclear weapons development under General Electric at sites (like Washington's Hanford Site).
69 'Hiroshima-Nagasaki Week', *Peace Newsletter* (August 1980), p. 5.
70 Graphic appears on the article by Geoffrey Navias, 'Three Movements: Strategies for the 1980's', *Peace Newsletter* (August 1980), p. 10.
71 Navias, 'Three Movements: Strategies for the 1980's', p. 10.
72 Malmsheimer, 'Three Mile Island', pp. 35–52 (p. 35).
73 Halevy, *Yes, I Glow in the Dark!*, LOC 353.
74 'A Cloud of Gloom and Doom', *New Scientist*, April 23, 1987, p. 17.
75 John A. Meyers, 'A Letter from the Publisher', *Time*, April 9, 1979, 113.15, p. 2.
76 Michael Swanwick, *In The Drift* (New York: Ace Science Fiction Books, 1985), p. 40.
77 Jean Baudrillard, 'The China Syndrome', in *Simulacra and Simulations*, by Jean Baudrillard, trans. by Shelia Faria Glaser (Michigan: University of Michigan Press, 2012), pp. 53–58 (p. 53).
78 Baudrillard, 'The China Syndrome', p. 57.
79 Baudrillard, 'The China Syndrome', p. 57. Italics in the original.

CHAPTER 5

Fears and Fallout

Three Mile Island's Legacy, Chernobyl, and Fukushima

When Homer Simpson (accidentally) prevented a full meltdown at the Springfield nuclear plant in an episode of *The Simpsons* in 1995, plant owner, Mr. Burns, praised the operator: "Homer, your bravery and quick thinking have turned a potential Chernobyl into a mere Three Mile Island."[1] However, there was nothing 'mere' about Three Mile Island. The 1979 event had a considerable impact on the nuclear industry and informed safety procedures for other plants. As Colin Tucker notes, "If you work at a PWR [pressurized water reactor], Three Mile Island is absolutely there."[2] The 1979 crisis also had a profound impact on American nuclear culture and public understanding and experience of nuclear power.[3] This is why Three Mile Island continues to surface in twenty-first century popular culture, even featuring in a multimedia opera.[4]

When faced with severe ruination and catastrophe resulting from later incidents like Chernobyl and Fukushima, there can be a tendency to view Three Mile Island as a mere blip that was easily resolved. To the contrary, as Daniel F. Ford points out, not only was the crisis "a complicated and protracted event" but it "tested a very large number of individuals, plant components, safety systems, instruments, communication links, emergency procedures, and much more".[5] Health studies, legal cases, cleanup, and industry and policy adjustment stretched well into the 1980s and beyond. When President Jimmy Carter left the plant on April 1, 1979 he did not take the crisis with him. When Thornburgh declared the crisis over on April 4 during NBC's *Today* show, the consequences did not simply vanish. As this book draws to a close, we will turn to look at some of the ways Three Mile Island still has influence, from official reports and changes in the industry, to more contemporary concerns over terrorism.

LESSONS FROM THREE MILE ISLAND

The biggest lesson to take from Three Mile Island is "accidents can happen".[6] This realization in an industry sold as peaceful and safe was greatly concerning. For William Freudenburg and Eugene Rosa, Three Mile Island reminded the public of the risk of nuclear technology. They explain that:

> No other single event, it is now clear, has led to a greater decline in support for peaceful use of the atom. Public confidence in the nation's nuclear program was shaken to the core. Shaken, too, was the future of nuclear power.[7]

For its role in the crisis, Metropolitan Edison was fined heavily for reported violations of Nuclear Regulatory Commission (NRC) regulations. Three Mile Island would not see the end of nuclear accidents, of course. As pointed out by Toby Moffett at the start of the report *Nuclear Safety—Three Years After Three Mile Island*, accidents continued at other plants, revealing "embarrassing design error and poor quality construction".[8] Three Mile Island did not only flag up issues for the future of the industry; it called into question decisions of the past. In hindsight, it was found that

> the rush to commercialize resulted in the early 1970's in inadequate safety reviews, an erratic licensing process, disregard for quality assurance, overblown need projections, and incorrect cost development, a premature commitment to reprocessing and breeder development, and inadequate attention to problems of environmental impact, weapons proliferation, and waste management.[9]

NRC Commissioner Peter Bradford best summarized the problems with the Commission that contributed to the crisis: "the NRC was taking much too much for granted in areas such as operator training and reactor instrumentation".[10] Both the Kemeny Report and Rogovin Report found functional problems with the way the NRC operated at the time and suggested substantial improvements and reorganization. Three Mile Island became a landmark event that enabled these issues to be highlighted and debated.

Phil Noble makes the crucial point that the Three Mile Island accident was not a display of recklessness, nonchalance, or even carelessness. He explains that the operators did what they could with the knowledge and training they had in that moment:

> They thought they were good, they thought they were the best, they followed what they were trained to do, they recognized the symptoms that caused them to take particular courses of action. With hindsight they were proved to be wrong, but at the time it wasn't the case that they were a bunch of cowboys. Absolutely not. They were trained to the highest standards available at the time and they did, in good faith, what they thought was the right thing to do.[11]

What the Three Mile Island incident led to was considerable industry shakeup and revisions with the NRC. The 1979 crisis had highlighted many issues that needed attention and, in many respects, the 'silver lining' of such a crisis was that improvements could be made to prevent a worse accident in the future; this is something Noble reflects on when he states that, although an accident is never a positive thing,

> the industry took a massive leap forward as a consequence of the lessons learned from Three Mile Island; if we hadn't have had that event we wouldn't be where we are now. We wouldn't have learned as much as we did.[12]

Focus fell to improving training (including further recruitment of highly skilled and educated workers)[13], tightening and reassessing procedures, and better governmental regulation (including regulating the design, operation, and the transportation and disposal of fuel). Attention was given to improving inspector programs at the NRC as well as strengthening how (and how often) plants were assessed.[14] Moreover, due to the struggle with communication, there was a call for improved emergency procedures and demand for proper reporting of errors or failures detected in plants. Concerning how the plant operated, reports called for improved control room design and operation (including better manning of night-time shifts), and refining the technology to prevent similar failings (such as improved reactor design, and continually inspecting and upgrading the technology).[15] Agreeing with the Kemeny Report, President Carter authorized many of the changes the commission suggested. Some of the first changes implemented involved changes to the NRC, notably the appointment of a new chairperson with greater powers and the placement of federal inspectors at all sites.[16] William Murray explains that over 100 improvements were implemented as a result of the accident and these included equipment upgrading, preventative maintenance, repairs, and "man/machine" improvements.[17] One of the most significant changes was the establishment of the self-regulating INPO (Institute of Nuclear Power

Operations) in 1979 in response to the Kemeny Report. Essentially, the INPO is focused on ensuring public safety and the safe, reliable performance of commercial plants.[18]

The Three Mile Island incident did not end in the summer of 1979.[19] As the *Washington Post* reported, Three Mile Island continued to provoke stress and anger especially when residents, returning from evacuation, discovered a decrease in house prices, and found that shops were promising not to sell produce from the area for those worried about purchasing contaminated milk; worse yet, residents faced the news that they would be paying towards cleanup costs.[20] The cleanup and decommissioning of Unit-2 was no small task and to complicate the situation, the NRC had not been faced with such an operation at that level before. Cleanup involved several stages (such as venting radiation, removing radioactive water, general decontamination, and the removal of the fuel) that would take years to achieve at great cost. During this process, several issues occurred to further concern the public; one major problem involved the venting of radiation. Radiation was so high in the reactor building that Metropolitan Edison wanted to vent the radiation into the atmosphere, but this decision was met with strong and angry public opposition; PANE even attempted to stop the action legally.[21] Although the venting was eventually permitted based on low impact to public health, the protest caused a delay of eight months. Another setback occurred when the people of Lancaster vehemently (and successfully) opposed the discharge of water produced from the Three Mile Island accident into the Susquehanna.[22] During the cleanup process, that lasted until 1993, claims were made (and substantiated by the NRC) that the cleanup job was being rushed and consequently regulations were not being followed.[23] It was not only the cleanup that caused consternation. There were accusations that while changes had been made, old habits seemed to be resurfacing; for example, the restart of Unit-1 was delayed not only due to ongoing legal issues but because of the detection of cheating in operator examinations at the plant and the discovery that Metropolitan Edison falsified information on coolant leaks before the 1979 accident.[24] More recently, reporters drew attention to a delay in communication during a minor radiation leak in 2009 at Three Mile Island.[25] But it was not just Three Mile Island that received close monitoring. Due to the 1979 crisis, all plants were under the spotlight. In 1987, this spotlight drew attention to sleeping (or "inattentive") licensed operators and supervisors in the control room at Pennsylvania's Peach Bottom Atomic Power Station, leading to an assessment that operators displayed "poor judgment and a negative attitude toward safety".[26] These findings led to a temporary suspension of operation (and arguably inspired many sleeping nuclear operator stories in *The Simpsons*).

In the 1980s, health studies concluded that no substantial health issues resulted from the accident (NRC, 1985). In the 1990s, further studies supported these initial claims and concluded that no ill health effects were caused by the incident (for example, the National Cancer Institute, Columbia University, the Pennsylvania Department of Health, and Three Mile Island's Public Health Fund report).[27] However, questions surrounding whether radiation releases during the Three Mile Island crisis were damaging to health, especially whether they have contributed to cancer levels in the community, remain under debate for scientists like Steve Wing (*et al.*), who suggest that radiation levels were substantially higher than official data claims.[28] Two thousand plaintiffs (approximately) from the local area were so convinced of radiation contamination that they launched a lawsuit. However, in 1996 these lawsuits were dismissed. Nevertheless, debates over radiation levels and potential contamination have bounced back and forth between scientists over the decades. Just one example can be seen with the disagreement between Wing (who supported the lawsuit and believes the reports of higher radiation levels than officially reported[29]) and Mervyn Susser (an epidemiologist, part of the Three Mile Island Public Health Fund, who argues that the radiation release was minimal, and that sickness symptoms could be stress related). Both respond to each other's analysis in a series of conflicting papers.[30] While the passage of time soothed fears for some, others remain concerned years later with some couples so anxious about health impacts that they opted not to have children.[31]

In 2010, Ralph De Santis from Exelon (the energy company that now operates Three Mile Island) spoke to Chris Knight of *The Patriot*. Reflecting on whether the Three Mile Island 1979 accident made the plant "self conscious", he replied affirmatively and spoke of the "humbling" event that remains in the minds of the plant and industry itself.[32] The relevance of the 1979 event continues today because the crisis became "the prototypic model" for learning how a severe accident, in which the ramifications of the crisis are unclear, can impact the immediate community and the wider country.[33] This is why Three Mile Island has also been referenced alongside very different technological accidents and environmental crises. For example, reporters during the Deepwater Horizon oil spill (2010, known as Gulf of Mexico oil spill) referred to Three Mile Island when demanding higher industrial standards, as they did after the 1979 incident. One article, entitled 'BP disaster echoes TMI incident—what have we learned?' noted that both crises involved human and machine error.[34]

Of course, the most enduring comparison is between Three Mile Island and other nuclear accidents—namely Chernobyl and Fukushima. While

Figure 5.1 Memorial sign in front of Three Mile Island
Photograph by Grace Holden (2015)

Three Mile Island illustrated the potential for disaster, and in many respects acted as a 'What If' event, Chernobyl and Fukushima would serve as the outcome of those questions. In terms of scale and devastation, there is no comparison between Three Mile Island and the later events of Chernobyl and Fukushima but these three events would become united in prose and news as a clustering of nuclear power disasters. Although other nuclear accidents, to various degrees, occurred before the 1979 incident, it was this one, more than any other, that would come to have such a significant impact on popular culture and on how future accidents were articulated. Three Mile Island was supposed to be, in many respects, the 'near miss' through which lessons would be learned to prevent future accidents. In this concluding chapter, we turn to look at what Harold Denton called the "three unhappy families"—Three Mile Island, Chernobyl, and Fukushima.[35]

Chernobyl

After Three Mile Island, Mitchell Rogovin (of the NRC's Rogovin Report) stated that "No one would question whether there will be another accident. It's merely a matter of when".[36] Rogovin was mainly talking about the state of the nuclear power industry in America, and indeed there were other accidents on American soil. However, it would be Chernobyl in 1986 that would see not just 'another accident' but a considerably worse one. Chernobyl would take much of the focus away from the American event, making it a notorious global catastrophe and demoting the Pennsylvanian accident to a more localized American event. While Ginzburg summarized the Three Mile Island crisis as "the prototypic model" for accidents with unknown ramifications, Chernobyl was the model for "actual and extensive harm" that affected the world, rather than one country.[37]

Like Three Mile Island, the accident in the Ukraine's Chernobyl Nuclear Power Plant (formerly of the Soviet Union) occurred in the early morning, during a slow shift. In the early morning of April 26, 1986, during a system test, the Chernobyl plant, positioned on the Pripyat River, experienced a series of grievous failures which led to explosions and fires. The intensity of the heat damaged the core and caused the fuel to melt.[38] The containment structure was damaged, and a reported (approximate) two hundred million curies of radiation were released into the atmosphere and spread throughout Europe. Unlike with Three Mile Island, Chernobyl saw a full meltdown. Many firefighters, military, and civil crews (known as liquidators) who attempted to contain the disaster in the immediate days of the catastrophe were killed or severely harmed by the radiation pouring from the plant. The aftermath saw permanent mass evacuations and the eventual decay and ruination of buildings which transformed a once thriving community into a derelict, and radiation drenched ghost-town.

Renowned Australian editorial cartoonist Pat Oliphant, in an inked sketch published on April 26, 1986, depicts the disaster at Chernobyl with a caption that suggests the Soviet Union would dismiss concerns and remind the media that nuclear accidents happen frequently in America.[39] The main reference here, of course, is to Three Mile Island. Oliphant is not the only person to link Three Mile Island with Chernobyl—a far worse accident. Why would Three Mile Island be referenced during such a catastrophic crisis? One of the biggest comparisons John F. Ahearne makes is that both the American and Soviet crises displayed operator and industry "complacency" and insufficient experience/training.[40] And again, problems were flagged with communications and reporting. Adding to the crisis was a sense that a cloud of secrecy had fallen over the situation. It did not help that information regarding the design and operation of the plant was hard to get, as mentioned in the United States hearing of the Chernobyl accident. This led to more confusion and suspicion as industry and government silence worsened an already tense situation (similar would occur during Fukushima). In fact, the secrecy over the accident in the

Soviet Union recalled not only the misdirection experienced during Three Mile Island but the secrecy with Windscale in 1957, and the surreptitious Urals disaster as well.[41]

Up to the present day, harrowing stories of health problems from the accident continue to be reported. A disturbing list of medical complaints and personal stories of ongoing tragedy is covered in the two documentaries offered by Maryann DeLeo, both of which explore the ruins of the Ukrainian disaster and the grave health consequences of the accident. In *Chernobyl Heart* (2003) DeLeo stands overlooking the reactor sarcophagus and is filled with emotion "to think of that innocuous little complex over there . . . that building has caused the destruction of nine million lives, half of which are children under the age of five".[42] Both DeLeo's *White Horse* and *Chernobyl Heart* deal with the ramifications of the tragedy on children and show that while time has passed, nuclear disasters linger through the impact they have had on the environment and community. In *White Horse*, Maxim Surkov, a former resident of the area, returns to the abandoned apartment building he lived in during the Chernobyl crisis twenty years earlier, which is now derelict and contaminated. Surkov finds an old calendar hanging on the wall and tears it so the year 1986 ceases in April. Surkov makes the point that the Ukraine disaster was so severe that for the hundreds of thousands of evacuees their world did end there. In Bill Toland and Clif Page's 'Ghosts of Chernobyl' (2002), the photographic report tells of how some people have chosen to return to their homes in the Dead Zone. Toland and Page recount the story of Olena Abrajez, who lives off the vegetables harvested from the irradiated land. The 'ghost' these people live with is the radiation that surrounds them every day, either due to obliviousness or disregard for the danger it poses. Poignantly, Toland and Page describe a picture of a dormant transformer yard in Chernobyl as "a futuristic Stonehenge".[43] Chernobyl, then, is a monument to our times: a present memorial to an ongoing disaster.

Fukushima

If the 9.0 magnitude earthquake and the thirty-foot tsunami were not enough, a nuclear meltdown on the International Nuclear Event Scale at 'Level 7' magnitude (the most severe) would become the third disaster for the Japanese in a series of incredibly tragic and unfortunate incidents.

On March 11, 2011, an earthquake caused the shutdown of reactor units 1, 2 and 3 of the Fukushima Daiichi plant operated by Tokyo Electric Power Company (TEPCO), and led to the severance of the plant from the nation's electricity grid.[44] The earthquake was damaging, but the following tsunami was crippling. The tsunami destroyed a vast array of buildings and services but most crucially wiped out the emergency generators, cooling pumps, electrics, and power supply, leaving the entire plant powerless.[45] These events not only meant extreme damage to the plant but the inability to cool the reactors quickly and safely to prevent a meltdown. In addition, the damage caused to the plant and surrounding area made it incredibly difficult for the disaster to be communicated. Hours later, at 7.30 p.m., the then Prime Minister Naoto Kan announced that a nuclear crisis was in progress; a mere hour and a half after this, an evacuation was ordered for all those within three kilometers of the plant. Like with Three Mile Island, the crisis began in the month of March, but the struggle to regain control of the plant would stagger through several months until shutdown was declared in December.

With the Fukushima disaster, Japan knew the ramifications of nuclear technology from both angles—as a weapon and as a dangerous nuclear power. Fukushima occurred during what some call 'the nuclear renaissance', a time in which renewed and enthusiastic attention returned to nuclear power as a greener solution than other energy sources. As Harry Henderson notes, the development of the plant also occurred after the Japanese public, who were initially resistant to nuclear power after Hiroshima and Nagasaki, had been convinced by the government of the merits of nuclear energy.[46]

The disaster at the Fukushima plant could not simply be dismissed as bad luck caused by extreme natural phenomenon. Kiyoshi Kurokawa, chairman of the Independent Investigation Commission, states that the nuclear crisis was not a product of a natural disaster. He argues that although the natural disaster was not preventable, it could have been contended with more effectively:

> The earthquake and tsunami of March 11, 2011 were natural disasters of a magnitude that shocked the entire world. Although triggered by these cataclysmic events, the subsequent accident at the Fukushima Daiichi Nuclear Power Plant cannot be regarded as a natural disaster. It was a profoundly manmade disaster—that could and should have been foreseen and prevented. And its

THE RESONANCE OF FICTION

In light of the Fukushima disaster, German broadcasters and other European countries such as Austria and Switzerland carefully vetted episodes of *The Simpsons* to avoid episodes that featured satirical nuclear crisis content.[47]

> effects could have been mitigated by a more effective human response.[48]

Kurokawa's report proceeds to cite numerous human errors and failings that significantly contributed to the events of March 2011. The report states that the Japanese nuclear industry had not learned from Three Mile Island and Chernobyl, which led to management error, operator error, secrecy, negligence, gross miscommunications, and a dangerous mix of "collusion" and "lack of governance" by "the government, the regulators and TEPCO".[49] Like with Three Mile Island, issues were found with training, regulation, operation, crisis management, and confused (and often subjective) information dissemination. The plant itself did not have the countermeasures in place to withstand the earthquake and tsunami, and the accident worsened due to insufficient backup systems and inadequate preventative measures.[50] Arnie Gundersen, at the World Uranium Symposium (2015), noted that the mistakes highlighted by Three Mile Island were not in fact learned from and that many mistakes were repeated at Chernobyl and Fukushima; in fact, the 2011 incident never should have occurred and could, and should have, been prevented.

With Three Mile Island, I previously commented on the distinction between technological and natural crises, specifically how the nuclear accident and the complex nature of potential ramifications (like radiation) caused a very different type of experience than with natural disasters which are often familiar, understood, predictable, and without accountability. However, Fukushima was unique as it enmeshed a technological disaster with natural disasters. During Fukushima, the technological aspect (such as concern over radiation) often overshadowed the natural dimension of the crisis (earthquake and tsunami). An example of this can be seen in the story of Sadako Shina published in the *New York Times* on March 17, 2011. Shina, who lived near the plant, refused to move from her property when both the earthquake and tsunami hit; however, when the nuclear crisis occurred, the fear of radiation made her leave the area.[51]

Matthew Lavine notes that while the events at Fukushima came as a surprise, the nuclear discourse that followed was familiar "because it had been rehearsed after Chernobyl, Three Mile Island, and dozens of lesser nuclear crises".[52] Charles D. Ferguson actually referred to Fukushima as a "movie" and remarked "We'd seen this movie twice", referring to Three Mile Island and Chernobyl.[53] There is a cultural history and cultural memory already in place and this is why the discourse is familiar. The familiarity of nuclear accident discourse can lead to the belief that all reactors are suspect, and crisis is repetitive. With nuclear crises specifically (although this is relevant to all technological crises), there is a pressing imperative to learn quickly from disaster and implement procedures to prevent a repeat scenario. Fukushima obviously caused America to reflect again on nuclear risk. Over thirty years after the Three Mile Island crisis, many questioned whether the ten-mile evacuation zone around Three Mile Island should be extended due to the twelve-mile evacuation issued by the Japanese government.[54] Those living near nuclear plants in areas of the United States that are susceptible to extreme weather had to be reassured that their local plants could withstand seismic activity.

"Three Unhappy Families" and the Media

In 1980, a report on the impact of Three Mile Island abroad commented that the 1979 accident highlighted numerous problems with communicating complex technological information to the media and that the need for communication can be invasive and demanding.[55] The NRC's Richard Vollmer stipulated that specialists must be trained in how to communicate with the press and public on technological and scientific stories of complexity.[56] In the report 'Staff Studies, Nuclear Accident and Recovery at Three Mile Island' (1980), recommendations included using one spokesperson to share NRC information and the establishment of a clear "voice of the agency" so that internal and external communications could be strengthened.[57] It was imagined that such implementations would lessen the crisis of miscommunication, contradiction, and perceived secrecy by authorities, as experienced during the Three Mile Island incident.

Unfortunately, during Chernobyl and Fukushima communication was worse than experienced in 1979. According to *The Bulletin of the Atomic Scientists*, after a dribbling of news following the accident on April 26, it was not until May 6 that a detailed report was issued in the Russian broadsheet *Pravda*.[58] By May 7, it was becoming apparent that earlier Soviet reports minimized the extent of the disaster. Conflicts emerged in the press regarding the differences between how the Soviet Union dealt with crisis compared to the West; it seemed to be that the West was more readily circulating information about the Chernobyl disaster while the Soviet Union tried to remain tight-lipped. An international media scuffle began, with the West seemingly pitted against the Soviet Union. Finally, on May 14, 1986, Soviet Leader Mikhail Gorbachev addressed his people and spoke about the accident in an announcement on the Soviet evening news. While Gorbachev addressed some details of the accident, he also attacked the Western media for "lies" and attempted to divert attention to nuclear weaponry (as being a greater risk). The conflict between the Soviet and Western media worsened at this point. ABC News used the Gorbachev speech to highlight more inconsistencies over the severity of the accident and also used the segment to point out that Chernobyl was far worse than Three Mile Island, praising American containment. However, one of the reasons Gorbachev cited for the delay in his speech to the people was that the leader wanted to wait until all the facts were known. Arguably, this is a reflection (and attack) on the confusion that plagued Three Mile Island.

Many American news publications were quick to challenge the Soviet reports and claimed, often without adequate evidence, exaggerated death tolls and conspiracy—the *New York Post* even claimed secret mass graves.[59] Many reports focused on the miscommunication and slowness of information distribution as being characteristically Soviet, neglecting to remember that the same issues of problematic and slow information release occurred at Three Mile Island. Suddenly nuclear 'secrecy' and 'mistrust' was positioned as intrinsically and fittingly a Soviet problem and part of a history of Soviet subterfuge. As William A. Dorman and Daniel Hirsch aptly note, some reports seemed biased towards representing the Soviets badly.[60] In the early days of American reporting, the message relayed was that Chernobyl was a tragedy but one that could not and would not be repeated in America. Such sentiment was not shared by all and in fact, Dorman and Hirsch, writing in *The Bulletin of the Atomic Scientists*, suggested that Cold War bias may have influenced the rush to cast the crisis as uniquely Soviet.[61] Such binary reporting (Soviet Chernobyl versus American Three Mile Island) was not just grounded in Cold War anti-Soviet mentality but was further complicated by the Cold War race for nuclear supremacy. For America, their nation had secured the weapon first, tested it first, deployed it first, while the Soviets, despite their advances, were perceived to be lagging behind. Thus, when America averted nuclear power crisis, and the Soviets descended into a flat-out catastrophe, another dimension was added to this narrative: only America could control the Nuclear Genie they created. Anyone can install a nuclear reactor and make it operational, but can they control it? America can, was the insinuation. Moreover, add this to the rocky nuclear treaty talks bubbling away in the background and we have a heated dynamic in which the Soviets were suddenly on compromised footing.

Similar occurred with Fukushima. Again, many publications reported that complete and accurate information was not being circulated. The *New York Times* was vocal in claiming that the Japanese government was withholding important information regarding radiation and the fate of workers at the plant. Thornburgh spoke of "déjà vu" when learning about the Fukushima crisis, especially concerning the slow and confused information dissemination to the public.[62] Once more, it was common for news reports to reassure the American people that an accident like Fukushima could not and would not occur on American soil, mainly due to faith in American nuclear superiority.[63] Ann Bisconti makes the point that after Three Mile Island, and even after Chernobyl, nuclear support rose after experiencing a brief dip. Arguably, this was due to the American public interpreting Three Mile Island as an example of crisis avoidance, and viewing the Soviet's Chernobyl crisis as further demonstrating how America can prevent disaster.[64] Fukushima, in the press, was also expressed as a Japanese failure that could not be repeated in America.

One dominant uniting concern for all accidents was the fear of radiation. Radiation, as with Three Mile Island, dominated both reporting on Chernobyl and reporting on Fukushima. However, due to the progression of time and improved education, how radiation was explored in the media between 1979 and 2011 was very different. Sharon Friedman, writing in the *Bulletin of the Atomic Scientists*, notes that Fukushima coverage directly responded to the suggestions for media content improvement after Three Mile Island. Reports were more detailed, more factual, and used a range of informational tools to provide clarity (such as graphs, tables, images, videos, timelines, graphics, and data).[65] Often, this information was available globally via the World Wide Web. While Three Mile Island and Chernobyl have online presence today, reporting was limited to television news and paper press at the time. Fukushima occurred during a different time of media engagement; not only were large news outlets able to disseminate news via their websites, but individuals were able to contribute to the news. The public could upload their videos on YouTube, provide live updates on Twitter, blog about their experiences on Tumblr and LiveJournal, share 'missing people' reports on Facebook, add to news stories in comments sections, and submit photos, film, and witness statements to news outlets digitally.

However, the 2011 crisis was still complicated by slow, vague, and inaccurate official reports. In part, this can be attributed to the difficulties in reporting two severe natural tragedies alongside the nuclear crisis. So, while Japan's crisis benefitted from media news avenues, this did not automatically equate to higher quality news reporting. Furthermore, as the official report on the Fukushima accident disclosed, the information

TWENTY-FIRST CENTURY NUCLEAR ENGAGEMENT

Downloadable applications can alert users to nuclear news; apps like *Nuclear Sites Free* (2012) and *Nuclear Plants* (2012) inform users how close they are to a nuclear station and where to evacuate. *Nuclear Power Plants* (2013) enables users to see where accidents have occurred on the INES scale and *Radiation Map Tracker* (2014) conveys information about radiation levels in areas subjected to nuclear tests or accidents.[66]

disseminated was patchy, vague, and did not reach all members of the public at the same time (if at all). The report notes significant time discrepancies in when people learned about the crisis in the community and poor information distribution about evacuation; in a resulting study, residents expressed mistrust of the government and TEPCO and demanded an in-depth investigation.[67] Even in 2011, a lot of the same mistakes from 1979 were witnessed.

Despite the disturbing history Japan has with the nuclear bomb in Hiroshima and Nagasaki, nuclear power has not been subject to the debate one might expect. Chim↑Pom countered this in 2011 by firmly linking the 1945 nuclear war tragedy to the 2011 nuclear power accident by extending the atomic bomb mural 'The Myth of Tomorrow' by Taro Okamoto (displayed in Shibuya Station) with scenes of the Fukushima disaster. A member of Chim↑Pom noted that the piece seemed incomplete; the group's addition (added to the mural without permission) seemed to finish Okamoto's nuclear criticism.[68] Chim↑Pom called their addition 'Level 7', which corresponds to the fact that Fukushima was a Level 7 accident on the INES scale. One month after the 2011 disaster, Chim↑Pom artists risked personal harm to travel to the plant and perform an artistic act that would encapsulate the relationship between the people, country, and nuclear threat. Overlooking the crippled plant, Chim↑Pom members used red spray paint on a white canvas to recreate the Japanese flag and then added three rectangles of red around the crimson circle, effectively transforming the flag into the nuclear symbol. The artists seem to suggest that the identity of Japan is now synonymous with the nuclear disaster.[69]

THE PENDULUM SWING: ANTI-NUCLEAR SENTIMENT AND INDUSTRY RECOVERY

Anti-nuclear vigils still occur on the anniversary of Three Mile Island and this anti-nuclear sentiment does have an impact, to some degree, on the industry. Although the industry was in decline before Three Mile Island, between 1979 and 2001 seventy-one orders for new nuclear plants were canceled.[70] Bonnie A. Osif, Anthony J. Baratta, and Thomas W. Conkling, of *TMI 25 Years Later*, note that Three Mile Island "accelerated" the cancellations and while it was not the sole cause, it did have a "substantial impact on the industry".[71] Murray argues that Three Mile Island reigns as a symbol for some of "the beginning of the decline of nuclear power in this country [America]".[72] Maybe so, but in reality, and more pressing for the people, is that Three Mile Island symbolized a decline in nuclear trust. It is easy to get lost in the drama and tragedy of Chernobyl and Fukushima, but Three Mile Island was, as Stephen Klaidman describes, an "epiphany of the nuclear age".[73] It was this accident, Klaidman rightly notes, that proved that the nuclear industry could face catastrophe; and it was this accident alone that showed a remarkable shift in public attitudes towards the industry.

In Chapter 3, we saw how Three Mile Island triggered anti-nuclear protests in Washington D.C. and New York City. The 1980s onwards would see anti-nuclear protests in America continue with demonstrations against the development of Black Fox (Oklahoma), the Vermont Yankee plant (Vermont), and San Onofre (California). Both Three Mile Island and Chernobyl impacted the life of the Shoreham Nuclear Power Plant initially owned by Long Island Lighting Company. The plant met opposition due to ineffectual evacuation plans and the ongoing anxiety over accidents. Ultimately, the plant's operation was blocked and costs (without return) rose beyond acceptable limits. As a last resort, the plant was sold to the State (the Long Island Power Authority) in 1992 for $1 for decommissioning. Even in the United Kingdom, the anti-nuclear protests that hampered Sizewell B nuclear plant after Three Mile Island and Chernobyl now significantly impact plans for Sizewell C following Fukushima. So strong was the backlash after Fukushima that when Japan's Ohi nuclear plant was shut down for a routine inspection, local protest in 2012 prevented the plant from restarting; Ohi was one of fifty-two out of fifty-four reactors in Japan that shut down following the 2011 accident.[74] And, in Germany, Fukushima became the final nail in the coffin as the country aims to phase out nuclear power.[75]

While many health studies report that public health and the environment were not notably impacted by the radiation releases from Three Mile Island, mistrust of the authorities and widely reported personal stories of human, animal, and environmental harm have challenged—even undermined—these reports. Consequently, (whether correctly or not) fears over potential harm to health and damage to the environment continue to resonate. However, the risk of terrorism is now a pressing twenty-first century concern that accompanies these two chief issues.

The ongoing health debate

Despite official reports articulating otherwise, many influential researchers, scientists, and doctors argue that radiation contamination from Three Mile Island was greater than reported. Harvey Wasserman argues that claims of no fatalities at Three Mile Island is a "lie".[76] Books like the anti-nuclear text *Killing Our Own: The Disaster of America's Experience with Atomic Radiation* (1982) deal not only with the nuclear armaments program but

The Return of Windscale (Sellafield)

In October 1957, Sir William Penney chaired the Board of Enquiry to compile a report for the Chairman of the United Kingdom Atomic Energy Authority (to be known as the Penney Report). A version of this report was submitted to Parliament in 1957, but the full report was only made available in 1988 through the Public Record Office. When privy to the full report, it became apparent that many details in 1957 had been stifled. In 1988, the *New York Times* ran a special reflecting on the Windscale accident and claiming that Prime Minister Harold Macmillan "suppressed" details of the Windscale accident over concerns that such details would harm the nuclear relationship between the United Kingdom and the United States.[77]

The plant would continue to be a source of controversy in the twenty-first century under its new name Sellafield. *The People* newspaper, in June 2003, reported on the strong community drive to have Sellafield permanently closed. The same publication noted demands from the British and Irish people that Brian Wilson (the then energy minister) not dismiss their concerns over cancer and other disorders including Down's syndrome that could be potentially linked to the Windscale accident.[78] The backlash was also reported by the *BBC* (May 30, 2003), which made reference to the 2003 report by the Irish Radiological Protection Institute which claimed that the primary source of contamination in the Irish Sea was Sellafield's radioactive waste discharge.[79]

with the nuclear power industry as well, from mining to waste. *Killing Our Own* dedicates two chapters to animal and human death as a result of Three Mile Island. Arnie Gundersen, a former reactor operator who managed projects at over seventy nuclear plants, is also very vocal about the risk nuclear power can bring. Citing Wing, Gundersen argues that many have died as a result of nuclear accidents, even at Three Mile Island. Gundersen further makes the case that, so far, the industry has seen an accident every seven years.[80] In 2009, Gundersen published a PowerPoint presentation in which he set out to dismiss the 'myths' of the crisis at Three Mile Island, especially with regard to accurate radiation levels. Gundersen, among others, suggests that radiation containment at Three Mile Island failed, and that radiation releases could have been considerably higher than the NRC reported.

Regardless of the level of radiation release, experts like physician and Nobel Peace Prize winner Helen Caldicott firmly note that radiation is unsafe regardless of the dose.[81] Caldicott is exceptionally outspoken in her opposition to nuclear power as can be seen in her successful book *Nuclear Power Is Not the Answer*. For Caldicott, Three Mile Island did cause deaths, and the death toll of Chernobyl exceeds 1 million; the fact this is not widely known, she suggests, is evidence of a "cover up".[82] Many anti-nuclear activist groups continue to articulate grave concerns over potential dangers facing the public and predict worse accidents to come; this is something Greenpeace suggests in 'An American Chernobyl' (2006) by noting almost 200 nuclear energy close calls in the United States since Chernobyl.[83] Dr. Brent Blackwelder, from Friends of the Earth, spoke out against the danger of the nuclear power industry and talked of a nuclear "roulette", making the claim (linking together Three Mile Island, Chernobyl, and Fukushima) that a plant meltdown could occur every three years statistically.[84]

Yet it is critical to remember that many studies (official but also independent) argue that Three Mile Island did not release dangerous levels of radiation. What is clear in this section is that the mistrust of the nuclear industry and authorities is so strong that people are prepared to challenge official findings and now have the ability to do so—something that arguably would not have occurred at this level decades before during the nuclear power heyday. Suspicion (founded or not), is hardly surprising when there is a proven history of suppressing nuclear information, as we have seen during Windscale.

A New Threat: Terrorism

In *The China Syndrome*, Jack Godell (Jack Lemmon) is able to invade the plant's control room, steal a gun from an armed guard, and remove all operators from the room at gunpoint. Even as a subplot, *The China Syndrome* highlights yet another risk posed by the nuclear industry—the potential for terrorism or sabotage.

Concerns over nuclear terrorism were actually present during the Atoms for Peace period. As we may remember, when pitched to the public, Atoms for Peace was seen as a concentrated effort to reposition nuclear technology as not only peaceful but *useful* to society. However, it was hard to liberate the atom from the shadow of the mushroom cloud, and many remained skeptical over the presence of atomic power in the form of power stations. One reason for this was the fear that sharing nuclear technology with other nations in the name of 'peace' would actually increase military usage unintentionally. Alan Cottrell makes a profound point when he states, "It is ironic that the same word, *power*, can signify either electrical power or political/military power, for these two meanings converge in a real and disturbing sense when the power is nuclear."[85] As evidenced in Jan Prawitz's essay, from the 1976 volume *Facing up to Nuclear Power*, concerns in society at that time surrounded the idea that a country with many nuclear power stations could develop a large stockpile of materials suitable for bomb construction or, worse, a terrorist organization may seize the material and use it to create weapons.[86] This is an issue Clyde W. Burleson commented on in 1980 when he claimed that even nations who had no interest in developing nuclear weapons now had the option to if they pursued nuclear power. In other words, 'Atoms for Peace' could lead to 'atoms for war'. Burleson added that "sooner or later some not-too-disciplined, over-egoed demagogue heading a one-man dictatorship is going to have the capacity to set off a nuclear explosion".[87] Concerns over a nuclear weapon in the hands of a dictator sparked the Manhattan Project initially, and here, in the 1970s and 1980s, thinkers such as Burleson reflected on Atoms for Peace and highlighted a similar concern. Prawitz noted that the attitude seemed to be that "peace and nuclear power are incompatible" and questioned whether it would ever be possible for nuclear power to coexist with peace.[88] In 1976, for Prawitz, there was no clear answer.

Today, when we think of terrorism we are likely to think about September 11, 2001. However, in 1993 an intruder hid inside the Three Mile Island complex for hours after invading the site by crashing through gates in a car. While the intruder did not apparently harbor terrorist inclinations, what became clear was that a member of the public was able to forcefully gain access to the plant. The terror attack on New York in 2001 worsened nuclear-terrorism concerns, and many questioned what could happen if a terrorist targeted a nuclear plant with a plane. In fact, just after September 11, it was claimed that al-Qaida had threatened Three Mile Island with a terror attack; it was even suggested that the United Airlines Flight 93 plane (which crashed in rural Pennsylvania) could have been initially intended to collide with Three Mile Island before going off course.[89] Consequently, Three Mile Island became the first plant to be tested in America for its provision against air strike based terrorism.[90]

Nearly all Three Mile Island texts considered in the last chapter reference terrorism to some extent. John Luciew's *Fatal Dead Lines*, which was inspired by the author's childhood experience of the partial meltdown, deals with a modern, revenge based terrorist attempt on the plant involving explosives. Luciew's book, in many ways, 'updates' the Three Mile Island crisis. In 1979, the threat was the breakdown of a 'mysterious' technology through machine failure and human error. The 1979 crisis also contained nebulous anxieties of the anti-nuclear movement, the nuclear element of the Cold War, and the psychological toll of Hiroshima and Nagasaki. However, when Luciew's novel was first published in 2003, concerns were not as much centered on internal technological issues as they were on the idea of remote warfare and terrorism.

After the September 11 terrorist attacks, Mohamed ElBaradei (the then Director of the International Atomic Energy Agency, under the United Nations) spoke of the assault as a forewarning that now requires the nuclear industry to ensure plants and nuclear materials are safeguarded against a similar attack.[91] In October 2001, George Bunn (consulting professor at Stanford University's Center for International Security and Cooperation) and Fritz Steinhausler (a visiting professor at the Center) addressed the issue of a terror attack on a nuclear power plant following the tragedy of September 11; they questioned what the ramifications would be if a hijacked plane was to career into a nuclear plant.[92] Bruce Williams (Exelon Vice President) explains that the 2001 terrorist attack changed how nuclear security is viewed and states that before the attack the idea of a plane intentionally colliding with a plant was not widely considered.[93] For Burns and Steinhausler, the attack on New York highlights that now is the time to concentrate on nuclear terrorism and nuclear sabotage.

In 1979, one of the most dominant concerns was about control: did America have control over its plants? Now the question is whether America has control over security. Fears of radiation silently creeping into the local area and harming innocents has now been joined by the fear that a terrorist may do the very same thing.

Ongoing environmental issues

As Harry Henderson neatly summarizes, the fear of invisible, insidious radiation made way for another invisible threat that received a lot of scientific, media, and public interest: greenhouse gases like carbon dioxide (CO_2).[94] Today, anxiety over greenhouse gases damaging the environment and contributing to global warming is more dominant, especially in media reporting, than nuclear radiation. Nuclear radiation is linked mostly to exceptional accidents, whereas global warming is a current, constant, and planet-wide issue. Although global warming is not a new concept, it has become a pressing concern of the increasingly environmentally conscious twenty-first century. While the position adopted by many is that nuclear power is clean and can respond to the problem of climate change because it is an example of low-carbon electricity, there is also a very strong demand that other energy options are

TERROR TEXTS

Even in *The Simpsons*, terrorism and the nuclear power industry has featured as a theme. Although, after September 11, Season 25 presented a terrorism plot in 'Homerland' (Part 1 and 2, 2013), nuclear terrorism was referenced in the first season with 'Crepes of Wrath' (1990).

explored. Caldicott, for example, makes the point that nuclear power is not only immensely costly, but the promises that nuclear energy is green and does not involve CO_2 emissions are not entirely true.[95]

One leading issue is the subject of nuclear waste—namely how to store and dispose of the toxic material. In a short interview with *LancasterOnline*, Suzanne Webster, a professor from England, claimed that ignoring the problems with nuclear waste disposal and presenting nuclear power as completely 'green' was "greenwash".[96] In the twenty-first century, art activism is reacting to this problem as seen in *Black Square XVII* by Taryn Simon. This piece is intended to be unveiled in the year 3015 and is an amalgamation of waste from nuclear, pharmaceutical, and chemical plants that is not safe to be showcased until the next millennium.[97] The drive of the project is to make people think about waste, recycling, and storage.

Even ideologically there seems to be a philosophy that 'natural' energy sources are not only more beneficial but are deemed to be more modern (and dare I say, trendy). Even in 1979, there were subtle references to the crudeness of nuclear power and the superiority of 'natural' power. For example, Exxon's solar power advertisements promised "Energy for a strong America". One Exxon advertisement was featured within the early pages of *Time's* 'Nuclear Nightmare' story on Three Mile Island (April 9, 1979). If solar power made America 'strong', nuclear power was seen to make it weak. Nuclear power seemed to be suddenly positioned as a crude, man-made, contaminating power, whereas solar power was presented as channeling the almost divine powers of the sun. After the 1979 accident, many people started to consider (or seriously revisit) energy options. Thornburgh, during the Hearing before the Committee on Environment and Public Works, found Three Mile Island had made him reconsider energy options: "A few nights ago I asked the engineering profession in our Commonwealth to help us develop safe and reliable alternate sources of energy for our citizens."[98] He questions where we would all be today if the money spent on nuclear power had been spent on "clean and efficient ways in which to use the gifts of coal and other resources".[99] However, that is not to say a coal accident cannot occur, as the diminished population of the Centralia ghost town will remind us. Other energy sources are frequently discussed, such as direct solar, satellite solar, wind and tidal power, wave power, ocean thermal energy conversion, hydroelectric power, biomass technology, geothermal technology, and so on.[100] But whether these methods would be able to generate the base load needed to live comfortably within our energy requirements is debatable, which renders nuclear power a 'necessary evil' for some. Eugeniya Stepnova, a doctor and professor who treats Chernobyl victims at the Ukrainian

Figure 5.2 Anti-nuke rally in Harrisburg calling for solar power
President's Commission on the Accident at Three Mile Island, March 29–April 30, 1979

Scientific Center of Radiology Medicine, echoes this point: "Accidents like Chernobyl are the price we pay for comfort. You can't expect people to simply stop using electricity. It is too late to put the genie back in the bottle."[101] This is not a new revelation. Mitchell Rogovin, in the aftermath of Three Mile Island, spoke of America as "hostages" to nuclear power.[102]

After 2011, a "tree hugging" sentiment was expressed by some, with a call for alternative and more 'natural' energy means to be explored.[103] This is an interesting perspective considering that an earthquake and tsunami contributed to the problems at the Fukushima plant; however, as we have seen throughout this book, there is a familiarity with nature and an understanding of nature that is not shared with technological crises. Nuclear power, which involves the intervention of man in nature through atom splitting, has been interpreted as the violent manipulation of matter. This, combined with the troubling issue of nuclear waste, and the despoliation of community areas and bodies of water (like the Susquehanna), has helped to demonize the industry. Nuclear technology, through its association with the atomic bombings of Japan, nuclear testing, and Plowshare, has seen the technology penetrating the earth, poisoning the environment, and sickening the people. However, nature, even though it can be violent, is viewed as something precious, idyllic, and preferable. Therefore, it is unsurprising that when a technological crisis occurs there is an instant call for these artificial machines to be retired in favor of advocating the 'God given' resource of the planet.

However, while anti-nuclear sentiment is still strong and attention turns to alternative energy means, the industry has many advocates. While Germany has turned its back on nuclear power, many countries are forging ahead with their nuclear-power programs. In a 2013 documentary ominously entitled *Pandora's Promise*, written and directed by Robert Stone, the nuclear power debate is examined with the concluding message that, for all its faults, nuclear power is the way forward. The title alone, with its obvious links to the famous Greek myth of Pandora's box (or jar), suggests—as did the myth—that humankind can inadvertently unleash evil upon the world through naive (or reckless) acts. However, the documentary focuses on the 'hope' that is trapped in Pandora's box: the 'promise' of nuclear power. The documentary concentrates on how nuclear sentiment has changed and interviews those who have made the move from nuclear opposition to nuclear support. This is not to say anti-nuclear protest is not still waging. In fact, the documentary opens with a rally over the Indian Point Energy Center in Buchanan, New York. One protestor introduces the documentary by claiming that nuclear power is "a bomb industry" and "wicked", and demands to know why President Barack Obama, a family man, supports such a perilous industry.[104] Nevertheless, the main thrust of Stone's exposition features interviews from individuals such as Steward Brand (founder of the Whole Earth Catalog), Pulitzer winner Richard Rhodes (author of *The Making of the Atomic Bomb*), Gwyneth Cravens (author of *Power to Save the World*), environmental activist Mark Lynas, and Michael Shellenberger (founder of The Breakthrough Institute)—people who have altered their stance on nuclear power and are now in favor of the promise offered by the technology. Although the documentary covers incidents such as Three Mile Island, Chernobyl, and Fukushima, all speakers articulate a shift from anti-nuclear sentiment to a pro-nuclear stance. The reason for this switch, for most, is the "quality of life" offered by constant power that comes without the burning of fossil fuels.[105] It is also important to note that leading climatologist James Hansen has argued for more focus on nuclear power to move the energy industry away from fossil fuels and CO_2 emissions. Hansen and Pushker A. Kharecha claimed in 2013 that nuclear power had prevented over a million air-pollution related deaths and could prevent millions more.[106]

The nuclear industry, despite four large-scale accidents, is pushing forward. Although nuclear plants are expensive to construct initially and there remain safety concerns over radiation and nuclear materials, there are numerous positives to nuclear power: the operating costs are low, fossil fuels are not used, and the uranium required is plentiful. Although Germany is winding down its nuclear industry, this is perhaps of no surprise considering that Germany has not been known as a pro-nuclear nation. While Sweden held a nuclear referendum after Three Mile Island in 1980, this did not mark a mass exodus from nuclear power. In fact, the vote to withdraw from nuclear power was overturned in 2009, showing again that times have changed. Although nuclear power is controversial in Sweden, as of 2016 there are plans to replace old reactors and Sweden is

Moving Out of Three Mile Island's Shadow? International Concerns: Sizewell B

It was a year after Chernobyl that the Central Electricity Generating Board was allowed to proceed with the construction of an American designed pressurized water reactor (PWR) in Suffolk, England. The reactor, to be known as Sizewell B, was described in the press as the same reactor responsible for the Three Mile Island crisis.[107] This concerned the local people for, as John Valentine remarks, the Three Mile Island accident "discredited the PWR technology".[108]

At this time, the United Kingdom's experience of nuclear power had been tainted by ongoing revelations of the Windscale accident. Furthermore, until this point, the country had been using advanced gas cooled reactors so the understanding the British public had about pressurized water reactors was limited to military submarines and Three Mile Island. Colin Tucker explains the concerns relating to the PWR design in the 1990s: "So, as a country, we are about to launch into the unknown with an American design . . . oh and, by the way, they had Three Mile Island. It's not surprising people were very, very nervous of them."[109] However, for the industry, Three Mile Island was one event amongst a whole fleet of safely functioning PWRs; further, what Three Mile Island showed the industry was that "even if you melt a third of the core, you can actually contain it".[110] The same understanding was not necessarily reflected in society.

Although closely associated with Three Mile Island, there are significant differences between the Sizewell B plant and the Pennsylvanian plant. Also, because Sizewell B was developed in response to the lessons learned during the 1979 crisis, it has benefitted from hardware design upgrades and plant operation improvements.[113] Recently the United Kingdom and EDF Energy have seen developments in the nuclear industry; while Sizewell C is still in consultation stages, in September 2016 Hinkley Point C was approved for construction in Somerset. EDF Energy offers tours of its plants and has opened many visitor centers, inviting community involvement. Although protests over new developments (Sizewell C and Hinkley Point C) have occurred, EDF's drive for public engagement has helped smooth the transition.

set to phase out the tax on nuclear plants to further assist the industry. Although Three Mile Island's operating license should have terminated in 2014, the NRC has prolonged the license (for Unit-1) until 2034. Surveys show that support for nuclear power and support for Three Mile Island is high in Pennsylvania.[111] One main reason nuclear power is supported is that many believe nuclear power is the only way to meet the increasing energy needs of the power-hungry twenty-first century. In America, nuclear "rejuvenation" can be seen through the restoration of Browns Ferry Nuclear Plant in the twenty-first century, decades after its notorious accident in 1975.[112]

WE ALL LIVE ON THREE MILE ISLAND

There has always been a magical wonder to nuclear technology; we saw this in the early days of radium and X-rays when optimism was high and the nuclear dream seemed limitless. From the magic of the atom came a technological sublimity (to recall David Nye), as the awesome spectacle of the atomic explosion was so magnificent and unparalleled that the mushroom cloud provoked both wonder and fear. Partly helped by Disney's personification of the nuclear through the humbling Nuclear Genie, there was most certainly a sense of magic and mystery associated with man's newest discovery. The sense of control America had over this new and mystic power, as demonstrated through the 1945 detonation in Japan, almost Americanized this 'fairytale'. Yet, the 'magic' did not, and could not, last.

In Chapter 1, we learned that apocalypse has shifted from religious connotations to a more secular understanding of disaster. Yet, there is more to it than that. In 'Of an Apocalyptic Tone Newly Adopted in Philosophy' (from a 1980 lecture), eminent philosopher Jacques Derrida explores the etymology of Apocalypse and traces the term back to the Hebrew *gala* meaning to reveal. Derrida explains that *gala* was translated into the Greek word *Apokalupsis/Apokaluptó* which means, "I disclose, I uncover, I unveil".[114] Similarly, David Herbert Lawrence in his reading of the Biblical 'Book of Revelation' states: "Apocalypse simply means revelation."[115] This revelation refers to the manifestation of Christ and the realization of God's will. Thus, the biblical Apocalypse demands both destruction and an unveiling. Thus, here we can say that apocalypse reveals something that was once hidden. Likewise, nuclear technology, after the early days of its magical fairytale narrative, now reveals something that was once hidden: radiation being one of the biggest 'invisible' threats. Ideas of a nuclear apocalypse need not refer to the nuclear holocaust of World War III, but can instead refer to the revelation of the dangers associated with nuclear technologies. It is with this revelation that the nuclear can bring forth disaster as seen with Chernobyl and Fukushima, that the magic of the peaceful atom is replaced with peril. Today we know about nuclear weapons and we know about nuclear power. We are aware of the benefits and the dangers, and we are wise enough to reflect that nuclear power was never going to be "too cheap to meter". The nuclear debate can no longer be aligned with a Disney-esque magic and mystery. There is no Nuclear Genie, and nor can we blame such a phenomenon when this personified entity gets out of hand due to human error and technological malfunction.

Throughout this book, we have seen the intimate connection nuclear weapons have with nuclear power. This is not to say they are comparable or even that such an association is helpful or accurate, but rather that—for better or worse—nuclear power and the Atoms for Peace program bloomed from that nuclear moment in 1945. Psychologically the links are absolutely there. Every time a nuclear accident occurs after Hiroshima there is an association (to some degree) with the 1945 tragedy because, in the words of Malmsheimer, there seems to be a "rematerialization of that threat".[116] Hence why protest signs of "Remember Harrisburg" are reminiscent of "Remember Hiroshima"; both events are linked in larger anti-nuclear rhetoric.[117] In Chapters 1 and 2, we explored how nuclear technology (whether weaponry or energy) has been hard to understand and how ambivalence has run high. There has been pressure to sway the American people towards or away from nuclear power using either extreme soothing or sensationalist terror. But anxiety over nuclear power is not just an offshoot of nuclear anxiety rooted in armaments; nuclear accidents have furthered the concern that the technology is not 'in hand' and challenged the promise of the peaceful atom. Distress experienced during Windscale, Three Mile Island, Chernobyl, and Fukushima is not merely a product of rearticulating World War II and Cold War nuclear hysteria; while this larger nuclear tapestry might inform such distress, these accidents stand for themselves. Those who fled Three Mile Island may have had Hiroshima, Nagasaki, nuclear testing, and radiation bubbling around in their consciousness, but it was that one plant in that specific month that provoked thousands of people to leave their homes. Nuclear accidents, like Three Mile Island, do not exist in a vacuum and this is what I have attempted to articulate throughout this book; however, at the same time these accidents do stand alone and are monuments to a very specific moment in time.

Today, it is still the case that many do not understand radiation and that any mention of it can cause panic. This is something Wade Allison, a nuclear and medical physicist, makes clear to the BBC by noting that far more had died in Japan from the 2011 tsunami than from radiation following the nuclear crisis. Allison argues for increased public education on what radiation is, how it works, and the dangers associated.[118] This is something Emma M. Cappelluzzo noted in 1979, just after Three Mile Island, when she stressed that schools need to educate on nuclear technology.[119] Cappelluzzo makes the point that technical terminology in the nuclear age can sound like a foreign language and during a crisis it is essential to educate the public on what is happening; but, beyond this, the American education system must educate teachers and students not just on the technology but on its cultural and contextual links. This will help the public assess a situation rather than react to it and will help the people take a stance on technology and form their own opinions.[120]

One of the reigning problems is that the industry and technology itself seem inscrutable, and it is difficult for the layperson to navigate the many issues of nuclear technology. Louise Bradford from Harrisburg makes the point in the *National Geographic* that following the Three Mile Island crisis, the people "have come to feel that they have too little control over their lives".[121] Feeling out of control is mainly due to not fully understanding the situation. The same article in the *National Geographic* describes the crippled cooling towers of Unit-2 as "silent giants".[122] This term is poignant because for so many people nuclear power is a giant and we can feel dwarfed by the plants, the industry, the policies, and the debates. Sure, the case can be made (and has been by thinkers like Brett McCollum[123]) that we face dangers every day, merely by crossing the road or consuming foods and drinks not healthy for us. However, this is not a reasonable comparison to make because daily risks (like driving a car) are chosen, comprehended, and limited. Nuclear power is a 'black box' to many people and when a crisis does occur it is rarely fully understood, and the ramifications seem sprawling. For McCollum, nuclear power is worth the risks it poses, and such risks will be limited by regulation and highly skilled, trustworthy operators.[124] But surely, one might argue, we thought we had good regulation and good operators in place during the last four major accidents. McCollum ends his article claiming that the risks that come with nuclear power are "clearly" worth it.[125] But, the point is that the risks are *not*, in fact, *clear* to many of us. Nor is nuclear technology in general *clearly* easy to understand.

Further, while McCollum may believe the risks posed by nuclear power is worth it, this is not a perspective shared by all. Therein lies a major problem. Where does the layperson turn? For example, do we look to Wing, Gundersen, and Caldicott and take the stance that, yes, Three Mile Island has significantly impacted human health and may continue to do so? Or, do we lean to the reports from the NRC and Kemeny Commission that radiation releases were not significant? How do we make this decision? The information is out there in abundance but for most it is difficult to know how to interpret this information. One of the biggest and most grievous assumptions made is that nuclear ambivalence is symptomatic of disinterest or nonchalance. Nuclear ambivalence, I suggest, is mostly rooted in not knowing how to work through the contradictory information we are presented with every day. A perfect example of this is beautifully presented in Henderson's book on nuclear power. In Henderson's text, two articles are side by side; one article is Brett McCollum's pro-nuclear piece in which the risks posed by the nuclear power industry are acceptable, while the other article is by Lloyd J. Dumas,' which concludes that the technology and industry are "too unforgiving", especially when combined with human operators who are susceptible to error.[126]

The only conclusion to reach, in the final moments of this book, is that Three Mile Island did have a considerable impact on the people of Pennsylvania but also in the wider world. America, the father of nuclear technology, had lost control of its own Nuclear Genie. And while Three Mile Island also reigns as an example of how America did contain the crisis and did prevent a catastrophe, the 1979 incident shook the world. In London, a 1979 issue of *The Economist* made the point that "We all live in Harrisburg" and remarked that the events of Three Mile Island would affect the citizens of the world, not only those who live close to plants.[127] Today, Three Mile Island is entwined with Windscale, Chernobyl, and Fukushima. The lumping together of nuclear crises as symptomatic of an entire industry is not difficult to believe considering that many do not have an in-depth understanding of the industry to differentiate. Murray notes that "[a] problem at one plant was now recognized as a matter for concern at all plants", leading to a "one world approach to nuclear safety".[128] Indeed, this is often the case. Technicians and physicists may roll their eyes over the meshing together of four distinct events but, for many, the nuclear power industry seems to be one body. As nuclear power is viewed as a global industry, a dramatic disaster can be felt everywhere. This is why after Fukushima so many news reports reflected on the American 1979 accident and contemporary nuclear sentiment in America. *LancasterOnline* ran the article 'Japan nuclear crisis stirs deep memories of TMI' (March 2011). Such sentiment was most poignantly expressed during Three Mile Island's thirty-second anniversary vigil in Middletown, Pennsylvania, during which some protesters held banners addressing the Japanese accident. One woman held a placard reading "T.M.I. Chernobyl. Fukushima. Who Is Next?"[129]

The 1945 bombing of Japan remains today a unique and isolated nuclear moment. Many do not envisage the repetition of such an event due to the 'promise' of mutually assured destruction. However, nuclear power accidents are shared and largely foreshadowed events that many can envisage reoccurring. Those countries that do have nuclear power will share, to some extent, the trauma of nuclear accidents abroad. Why was Three Mile Island such a rupture moment beyond American soil? Because it is relatable to all of us. As Bill Moyers for *CBS Evening News* said, "We all live on Three Mile Island."[130]

NOTES

1 'King Size Homer', *The Simpsons*, Fox, November 5, 1995.
2 Colin Tucker, interview with Grace Halden, Sizewell B, April 13, 2015.
3 Even in the twenty-first century, the Three Mile Island crisis is remembered during anniversaries; *The Record-Argus*, for example, remembered Three Mile Island separately in 'Today's History' and 'Today's Fact' on March 28, 2007.
4 The opera was created collectively by Karl Hoffmann, Guido Barbieri, Andrea Molino, and Oscar Pizzo after Fukushima (2012). The Three Mile Island Opera is available (with English subtitles) to watch on the Fairewinds website: Maggie Gundersen, 'TMI: A Human Perspective', *Fairewinds* (April 2, 2015), available online at www.fairewinds.org/demystify/three-mile-island-opera?rq=Three%20mile%20island%20opera (accessed August 2, 2016).
5 Daniel F. Ford, *Three Mile Island: Thirty Minutes to Meltdown* (New York: Viking Press, 1982), p. 253.
6 *Nuclear Powerplant Safety After Three Mile Island*, 'Report Prepared by the Subcommittee on Energy Research and Production of the Committee on Science and Technology', United States House of Representatives, Ninety-Sixth Congress, Second Session (Washington D.C., March 1980), p. 35.
7 *Public Reactions to Nuclear Power*, ed. by William R. Freudenburg and Eugene A. Rosa (Colorado: Westview Press, 1984), p. 24.
8 *Nuclear Safety—Three Years After Three Mile Island*, 'Joint Hearing before the Subcommittees of the Committees on Government Operations and Interior and Insular Affairs', House of Representatives, Ninety-Seventh Congress, Second Session (Washington, D.C., March 12, 1982), p. 1.
9 Peter A. Bradford, in *Nuclear Safety—Three Years After Three Mile Island*, 'Joint Hearing before the Subcommittees of the Committees on Government Operations and Interior and Insular Affairs', House of Representatives, Ninety-Seventh Congress, Second Session (Washington, D.C., March 12, 1982), p. 5.
10 Peter A. Bradford, in *Nuclear Safety—Three Years After Three Mile Island*, 'Joint Hearing before the Subcommittees of the Committees on Government Operations and Interior and Insular Affairs', p. 25.
11 Phil Noble, interview with Grace Halden, EDF Energy Nuclear, June 2, 2015.
12 Noble, interview.

13 Phil Noble, Head of Operations Development at EDF, worked at Three Mile Island years after the accident and reflected that the training he went through to be a licensed operator in the control room was rigorous, thorough, and extremely advanced. Noble commented that the heightened training was a response to the Three Mile Island incident of 1979 and that many of the training procedures were brought over to England: "it opened our eyes to what could be done". Noble, interview.

14 Susie Derkins, *The Meltdown at Three Mile Island* (New York: Rosen Publishing Group, 2003), p. 35.

15 See also *Staff Studies, Nuclear Accident and Recovery at Three Mile Island*, 'Subcommittee on Nuclear Regulation for the Committee on Environment and Public Works', United States Senate, Ninety-Sixth Congress, Second Session (Washington D.C.,: July 1980).

16 For a more detailed breakdown see 'Carter Acts on TMI: A Flood of Changes', *Science News*, 116.24 (December 15, 1979), pp. 405–406

17 William Murray, *Nuclear Turnaround: Recovery from Three Mile Island and the Lessons for the Future of Nuclear Power* (n.p.: 1st Books Library, 2003), pp. 30–31. A good overview is offered here: United States Nuclear Regulatory Commission, 'Backgrounder: Three Mile Island Accident' (2013), available online at www.nrc.gov/reading-rm/doc-collections/fact-sheets/3mile-isle.pdf (accessed August 2, 2016).

18 INPO, 'Home', *INPO* (2015), available online at www.inpo.info/Index.html (accessed 18 July 2016).

19 For detailed timeline of Three Mile Island developments in the 1980s see William Murray, *Nuclear Turnaround: Recovery from Three Mile Island and the Lessons for the Future of Nuclear Power* (n.p.: 1st Books Library, 2003)

20 'Inhabitants Wonder What to Believe', *WashingtonPost.com* (1979), available online at www.washingtonpost.com/wp-srv/national/longterm/tmi/stories/ch14.htm (accessed July 27, 2016).

21 Bonnie A. Osif, Anthony J. Baratta and Thomas W. Conkling, *TMI 25 Years Later: The Three Mile Island Power Plant Accident and Its Impact* (Pennsylvania: The Pennsylvania State University Press, 2004), pp. 34–35.

22 Noted in J. Samuel Walker, *Three Mile Island: A Nuclear Crisis in Historical Perspective* (London: University of California Press, 2004), p. 225.

23 See Eliot Marshall, 'Investigation Confirms TMI Cleanup Problems', *Science*, New Series, 221.4618 (September 30, 1983), p. 1357.

24 Detailed in Walker, *Three Mile Island*, p. 233.

25 See The Inquirer Editorial Board, 'How not to build trust in nuclear power', *The Inquirer Daily News* (November 27, 2009), available online at www.philly.com/philly/blogs/inq_ed_board/How_not_to_build_trust_in_nuclear_power.html?c=r (accessed July 22, 2016). The NRC held an 'open house' meeting to discuss with the public the plant's performance. See Dan Miller, 'NRC Tuesday to Hold Open House Regarding Three Mile Island in Londonderry Township', *The Patriot* (April 7, 2010), available online at www.pennlive.com/midstate/index.ssf/2010/04/nrc_tuesday_to_hold_open_house.html (accessed June 19 2016).

26 Order Suspending Power Operation and Order to Show Cause (Effective Immediately), United States Nuclear Regulatory Commission (Washington D.C., 31 March 1987), p. 3

27 Murray, *Nuclear Turnaround*, p. 134. See Maureen C. Hatch, *et al.*, 'Cancer Near the Three Mile Island Nuclear Plant: Radiation Emissions', *Journal of Epidemiology*, 132.3 (September 1990), 397–412; Evelyn O. Talbott, *et al.*, 'Mortality among the Residents of the Three Mile Island Accident Area: 1979–1992', *Environmental Health Perspectives*, 108.6 (June 2000), 545–552; Evelyn O. Talbott, *et al.*, "Long-Term Follow-Up of the Residents of the Three Mile Island Accident Area: 1979–1988", *Environmental Health Perspectives* 111.3 (March 2003), 341–348.

28 Steve Wing, *et al.*, 'A Reevaluation of Cancer Incidence near the Three Mile Island Nuclear Plant: The Collision of Evidence and Assumptions', *Environmental Health Perspectives*, 105.1 (January 1997), 52–57.

29 Steve Wing, 'Objectivity and Ethics in Environmental Health Science', *Environmental Health Perspectives*, 111.14 (November 2003), 1809–1818 (p. 1810). Wing was "impressed" with the work of Marjorie and Norman Aamodt (1984) who argued that radiation was considerably higher than reported.

30 Two illustrative samples from the same year, 1997: Wing, 'A Reevaluation of Cancer Incidence near the Three Mile Island Nuclear Plant', pp. 52–57; Mervyn Susser, 'Three Mile Island Accident Continued: Further Comment', *Environmental Health Perspectives*, 105.6 (June 1997), 566–567.

31 Reported by Thomas Bailey in Rich Kirkpatrick, 'Concerns Linger Ten Years After Three Mile Island', *The Dispatch*, March 25, 1989, p. 9.

32 Chris Knight, 'Three Mile Island Tour', *The Patriot* (8 September 2010), available online at http://videos.pennlive.com/patriot-news/2009/10/three_mile_island_tour.html (accessed August 5, 2015).

33 Harold M. Ginzburg, 'The Psychological Consequences of the Chernobyl Accident: Findings from the International Atomic Energy Agency Study', *Public Health Reports*, 108.2 (March–April 1993), 184–192 (p. 184).

34 Gordon Tomb, 'BP Disaster Echoes TMI Incident—What Have We Learned?', *PennLive* (June 20, 2010), available online at www.pennlive.com/editorials/index.ssf/2010/06/bp_diaster_echoes_tmi_incident.html (accessed June 12, 2016). April 20, 2010, an explosion on the Deepwater Horizon oil rig working for BP caused the death of 11 employees. As oil spilled into the ocean, aquatic life and bird life faced heavy death tolls making this accident one of the worst environmental tragedies. The disaster also greatly impacted local communities, jobs, tourism, and the economy.

35 Harold Denton talk in 'Nuclear Power Accidents', *Knox County Public Library* (April 29, 2011), available online at www.knoxlib.org/about/news-and-publications/podcasts/brown-bag-green-book-podcast/nuclear-power-plant-accidents (accessed July 30, 2016).

36 Quoted in: Daniel F. Ford, *Three Mile Island: Thirty Minutes to Meltdown* (New York: Viking Press, 1982), p. 261.

37 Ginzburg, 'The Psychological Consequences of the Chernobyl Accident', pp. 184–192 (p. 184).

38 See James A. McClure, *et al.*, *The Chernobyl Accident*, 'Hearing before the Committee on Energy and Natural Resources', United States Senate, Ninety-Ninth Congress, Second Session on the Chernobyl Accident and Implementations for the Domestic Nuclear Industry (Washington D.C., June 19, 1986), p. 51.

39 Pat Oliphant, 'Chernobyl!!', April 29, 1986, in Pat Oliphant and Susan Conway, *The New World Order in Drawing and Sculpture 1983–1993* (Kansas: Andrews and McMeel, 1994), p. 26.
40 John F. Ahearne, 'Nuclear Power after Chernobyl', *Science,* New Series, 236.4802 (May 8, 1987), 673–679.
41 *Nuclear Power in Crisis*, ed. by Andrew Blowers and David Pepper (London: Croom, 1987), p. 19.
42 *Chernobyl Heart,* dir. by Maryann DeLeo (HBO, 2003)
43 Bill Toland and Clif Page, 'Ghosts of Chernobyl', *The Times* (December 15, 2002) p. E4.
44 'Timeline: Japan Power Plant Crisis', *BBC* (March 13, 2011), available online at www.bbc.co.uk/news/science-environment-12722719 (accessed April 19, 2016).
45 For more detail see Kiyoshi Kurokawa, *et al.*, *The National Diet of Japan. The Official Report of The Fukushima Nuclear Accident Independent Investigation Commission,* The Fukushima Nuclear Accident Independent Investigation Commission (2012), available online at www.nirs.org/fukushima/naiic_report.pdf (accessed August 14, 2016), pp. 12–15.
46 Harry Henderson, *Nuclear Power: A Reference Handbook,* 2nd edn (California: ABC-CLIO, 2014), p. 46.
47 Example see: 'D'oh! German Broadcaster Pulls "Nuclear" Simpsons Episodes', *New York Post* (March 28, 2011), available online at http://nypost.com/2011/03/28/doh-german-broadcaster-pulls-nuclear-simpsons-episodes/ (accessed May 16, 2016).
48 Kurokawa, *et al.*, *The National Diet of Japan*, p. 9.
49 Kurokawa, *et al.*, *The National Diet of Japan*, p. 16.
50 See Kurokawa, *et al.*, *The National Diet of Japan*, pp. 30–32.
51 Martin Fackler, 'Radiation Fears and Distrust Push Thousands From Homes', *The New York Times* (March 17, 2011), available online at www.nytimes.com/2011/03/18/world/asia/18displaced.html?pagewante%20d=all%3E&_r=0 (accessed July 18, 2016).
52 Matthew Lavine, *The First Atomic Age. Scientists, Radiations, and the American Public 1895–1945* (New York: Palgrave Macmillan, 2013), p. 2.
53 Charles D. Ferguson, 'Japan Melted Down, but that Doesn't Mean the end of the Atomic Age', *Foreign Policy*, 189 (November 2011), 49–53 (p. 49).
54 Example article: Jeff Frantz, 'Experts Debate Three Mile Island's Safety Buffer', *Pennlive*, (March 19, 2011), available online at www.pennlive.com/midstate/index.ssf/2011/03/experts_debate_three_mile_isla.html (accessed November 11, 2015).
55 *Impact Abroad of the Accident at the Three Mile Island Nuclear Power Plant*, 'Subcommittee on Energy, Nuclear Proliferation and Federal Services of the Committee on Governmental Affairs', United States Senate (Washington: Library of Congress, May 1980), p. 2.
56 Richard Vollmer, 'Representing the Nuclear Regulatory Commission', in *The Three Mile Island Nuclear Accident: Lessons and Implications*, ed. by Thomas H. Moss and David L. Sills (New York: The New York Academy of Sciences, 1981), pp. 110–113 (p. 111).
57 *Staff Studies, Nuclear Accident and Recovery at Three Mile Island*, 'Subcommittee on Nuclear Regulation for the Committee on Environment and Public Works',

United States Senate, Ninety-Sixth Congress, Second Session (Washington D.C., July 1980), p. 34.

58 Alexander Amerisov, 'Chernobyl Emerging Story: A Chronology of Soviet Media Coverage', *Bulletin of the Atomic Scientists*, 3.1 (September 1986), 38–39 (p. 38).

59 Reported in William A. Dorman and Daniel Hirsch, 'The U.S. Media's Slant', in *Chernobyl: The Emerging Story, The Bulletin of the Atomic Scientists*, ed. by Harrison Brown (August/September 1986), 54–56 (p. 54).

60 Dorman and Hirsch, 'The U.S. Media's Slant', pp. 54–56 (p. 56)

61 Dorman and Hirsch, 'The U.S. Media's Slant', pp. 54–56.

62 Reported in Donald Gilliland, 'For Some, Japanese Nuclear Emergency Mirrors 1979's Three Mile Island', *Pennlive* (March 12, 2011), available online at www.pennlive.com/midstate/index.ssf/2011/03/for_some_japanese_nuclear_emer.html (accessed May 17, 2015).

63 This is something Arnie Gundersen also notes in his work on Fukushima, that Japanese and American systems are not radically different. See Arnie Gundersen, 'Fukushima', *Fairewinds* (n.d.), available online at www.fairewinds.org/fukushima (accessed July 20, 2016).

64 Ann Bisconti, 'United States Public Opinion About Nuclear Energy: Past, Present and Future', *From Its Birthplace: A Symposium on the Future of Nuclear Power* (University of Pittsburgh, March 27–28, 2012), available online at www.thornburghforum.pitt.edu/sites/default/files/Nuclear%20Symposium%20report%20FINAL%20report%2011_5_12.pdf (accessed May 22, 2016), pp. 37–38 (p. 37).

65 Sharon M. Friedman, 'Three Mile Island, Chernobyl, and Fukushima: An Analysis of Traditional and New Media Coverage of Nuclear Accidents and Radiation', *Bulletin of the Atomic Scientists*, 67.5 (September 2011), 55–65 (p. 60).

66 *Nuclear Plants*, developed by Claus Zimmerman, October 22, 2012, version 1.3. *Nuclear Power Plants*, developed by Michael Hoereth, October 10, 2013, version 2.7. *Nuclear Sites Free*, developed by Janak Shah, March 29, 2012, version 1.3.1. *Nuclear Tycoon*, developed by Justin Lehmann, May 17, 2016, version 1.0.1. *Radiation Map Tracker*, developed by Black Cat Systems, July 1, 2014, version 1.2.

67 See Kurokawa *et al.*, *The National Diet of Japan*, p. 60.

68 The artistic work of Chim↑Pom was covered by PBS. See 'The Atomic Artists', dir. by Emily Taguchi (PBS, 2011), in *Frontline* (2014), available online at www.pbs.org/wgbh/pages/frontline/the-atomic-artists/?autoplay (accessed May 25, 2016).

69 Many protest pieces surfaced after 2011 in Japan. For example, music artist Kazuyoshi Saito attacked the government for lies over the safety of nuclear power and attacked nuclear ambivalence in a musical piece. An anti-nuclear pop song by Seihuku Kojo Iinkai translated as 'Free from Nuclear Power Plant', also spoke out against the nuclear industry.

70 Osif, Baratta and Conkling, *TMI 25 Years Later*, pp. 85.

71 Osif, Baratta and Conkling, *TMI 25 Years Later*, pp. 85.

72 Murray, *Nuclear Turnaround*, p. 17.

73 Stephen Klaidman, *Health in the Headlines* (New York: Oxford University Press, 1991), p. 83.

74 See Martin Fackler, 'Japan's Nuclear Energy Industry Nears Shutdown, at Least for Now', *New York Times*, (March 8, 2012), available online at www.nytimes.com/2012/03/09/world/asia/japan-shutting-down-its-nuclear-power-industry.html?_r=2&emc=tnt&tntemail1=y (accessed July 17, 2016).

75 Note: Germany intended to phase out nuclear power before Fukushima. The 2011 disaster seemed to evidence why such a move was crucial.
76 Harvey Wasserman, 'People Were Killed by Three Mile Island and Other Nuclear Disasters', *OpEdNews* (November 19, 2016), available online at www.opednews.com/articles/genera_harvey_w_071119_people_were_killed_b.htm (accessed April 13, 2016).
77 Steve Lohr, 'Britain Suppressed Details of '57 Atomic Disaster', *The New York Times* (January 2, 1988), available online at www.nytimes.com/1988/01/02/world/britain-suppressed-details-of-57-atomic-disaster.html (accessed August 30, 2016)
78 'Voice of the People: A Sickening Way to Belittle Fears over Sellafield', *The People*, June 1, 2003, p.8.
79 'Sellafield Comment "Offensive" ', *BBC* (May 30, 2003), available online at http://news.bbc.co.uk/1/hi/northern_ireland/2949334.stm (accessed August 30, 2016).
80 Arnie Gundersen, 'Nuclear Science Guy Postulates Meltdown Every Seven Years', *HuntingtonNews.net* (February 9, 2015), available online at www.huntingtonnews.net/106198 (accessed March 18, 2016).
81 Speaking about Fukushima but the sentiment is true of all radiation releases. Helen Caldicott, 'Unsafe at Any Dose', *The New York Times*, April 30, 2011, available online at www.nytimes.com/2011/05/01/opinion/01caldicott.html?_r=0 (accessed July 21, 2016) p. 2.
82 Lecture: Dr. Helen Caldicott, 'Dr. Helen Caldicott: What We Learned from Fukushima', Seattle Community Media (2012), available online at http://seattlecommunitymedia.org/node/30758 (accessed July 17, 2016). Caldicott is referring to the claims in *Chernobyl: Consequences of the Catastrophe for People and the Environment*, which is a translation of a Russian text by Alexey V. Yablokov, Vassily B. Nesterenko, and Alexey V. Nesterenko, edited by Janette D. Sherman-Nevinger and reproduced for the Annals of the New York Academy of Sciences, Volume 1181.
83 Greenpeace, 'An American Chernobyl: Nuclear "Near Misses" at U.S. Reactors Since 1986', *Greenpeace.org* (2006), available online at www.greenpeace.org/wp-content/uploads/legacy/Global/usa/report/2007/9/an-american-chernobyl-nuclear.pdf (accessed July 30, 2016), pp. 4, 21.
84 Brent Blackwelder, 'American Forum: Nuclear Reactor Roulette', *The Phoenix News* (April 30, 2011), available online at www.phoenixvillenews.com/article/20110430/TMP06/304309971 (accessed May 12, 2015). Many documentaries have concentrated on the human element of disaster, especially the health element. The artistic poetry based piece *Heavy Water: A Film for Chernobyl* (2006) focused on the aftermath of Chernobyl and contended with the idea of home and community by visiting the abandoned city and locals who continue to live in the contamination zone. In regards to Fukushima, a moving example from 2013 is *Surviving the Tsunami: My Atomic Aunt* which follows families living close to the radiation zone and the impact the crisis had on the lives of locals and Japanese culture, as well as on the environment and economic situation in Japan.
85 Alan Cottrell, *How Safe is Nuclear Energy* (London: Heinemann, 1981), p. 102.
86 Jan Prawitz, 'Is Nuclear Power Compatible with Peace? The Relation between Nuclear Energy and Nuclear Weapons', in *Facing up to Nuclear Power*, ed. by John Francis and Paul Abrecht (Philadelphia: The Westminster Press, 1976), pp. 102–109 (p. 102).

87 Clyde W. Burleson, *The Day the Bomb Fell: True Stories of the Nuclear Age* (London: Sphere, 1980), p. 99.

88 Jan Prawitz, 'Is Nuclear Power Compatible with Peace? The Relation between Nuclear Energy and Nuclear Weapons', in *Facing up to Nuclear Power*, ed. by Francis and Abrecht, pp. 102–109 (p. 102).

89 Ad Crable, 'TMI 25 Years After: Security Now Top Worry', *LancasterOnline* (March 19, 2004), available online at http://lancasteronline.com/news/tmi-years-after-security-now-top-worry/article_563825a4-a986–561c-ba87-d0f38ab88844.html (accessed March 18, 2016)

90 See David Wenner, 'TMI will be First U.S. Power Plant Formally Tested on Terrorist Attack Plan', *PennLive* (April 16, 2013), available online at www.pennlive.com/midstate/index.ssf/2013/04/tmi_three_mile_island_drill_mi.html (accessed July 15, 2016)

91 'IAEA General Conference Adopts Resolution on the Physical Protection of Nuclear Material and Nuclear Facilities', *IAEA: International Atomic Energy Agency* (September 21, 2001), available online at www.iaea.org/newscenter/pressreleases/iaea-general-conference-adopts-resolution-physical-protection-nuclear-material-and-nuclear-facilities (accessed July 20, 2016)

92 George Bunn and Fritz Steinhausler, 'Guarding Nuclear Reactors and Material from Terrorists and Thieves', *Arms Control Today*, 31.8 (October 2001), 8–12 (p. 8).

93 Rebecca J. Ritzel, 'TMI a Look Back', *LancasterOnline* (March 24, 2004), available online at http://lancasteronline.com/news/tmi-a-look-back/article_2e446a14-d95e-53a4-959b-02eb042ab70f.html (accessed April 20, 2016)

94 Henderson, *Nuclear Power: A Reference Handbook*, p. 109.

95 Lecture: Dr. Helen Caldicott, 'Dr. Helen Caldicott: What We Learned from Fukushima', Seattle Community Media (2012), available online at http://seattlecommunitymedia.org/node/30758 (accessed July 17, 2016).

96 Amanda Balionis and Paul Franz, 'TALKING POINTS: More Nukes?', *LancasterOnline* (April 12, 2009), available online at http://lancasteronline.com/news/talking-points-more-nukes/article_8f797de0-0f4f-58b0-85b2-aaf9fa47d01c.html (accessed April 19, 2016).

97 Information on *Black Square XVII* here: Marina Garcia-Vasquez, 'Nuclear Waste Is Art in the Work of Taryn Simon', *The Creators Project* (June 9, 2015), available online at http://thecreatorsproject.vice.com/en_uk/blog/nuclear-waste-is-art-in-the-work-of-taryn-simon (accessed May 17, 2016)

98 *Three Mile Island Nuclear Powerplant Accident*, 'Hearings before the Subcommittee on Nuclear Regulation of the Committee on Environment and Public Works', United States Senate, Ninety-Sixth Congress, First Session (Washington D.C., April 10, 28, and 30, 1979), Part 1, p. 241.

99 *Three Mile Island Nuclear Powerplant Accident*, 'Hearings before the Subcommittee on Nuclear Regulation of the Committee on Environment and Public Works', Part 1, p. 241.

100 It is beyond the aims of this book to discuss the pros and cons of these sources. Please see Sam H. Schurr, *et al.*, *Energy in America's Future* (London: Johns Hopkins University Press, 1979)

101 Toland and Page, 'Ghosts of Chernobyl', p. E11.

102 Quoted in Daniel F. Ford, *Three Mile Island: Thirty Minutes to Meltdown* (New York: Viking Press, 1982), p. 261.
103 Clifford A. Rieders, 'Conservation: Maybe the Tree Huggers Were Right', *Phoenix News* (March 26, 2011), available online at www.phoenixvillenews.com/article/20110326/TMP06/303269970 (accessed April 17, 2016)
104 In *Pandora's Promise*, dir. by Robert Stone (Impact Partners, 2013)
105 Quote from Gwyneth Cravens, in *Pandora's Promise*, dir. by Robert Stone (Impact Partners, 2013)
106 Pushker A. Kharecha and James E. Hansen, 'Prevented Mortality and Greenhouse Gas Emissions from Historical and Projected Nuclear Power', *Environmental, Science & Technology*, 47.9 (2013), 4889–4895.
107 See David Dickson, 'Britain Chooses U.S.-Designed Reactor', *Science,* New Series, 235.4789 (6 February 1987), p. 629
108 John Valentine, *Atomic Crossroads: Before and After Sizewell* (London: Merlin Press, 1985), p. 43.
109 Colin Tucker interview with Grace Halden, Sizewell B, April 13, 2015.
110 Tucker interview with Grace Halden.
111 Survey 2008 by Susquehanna Polling and Research for Exelon. Garry Lenton, 'Poll Shows Midstate Supports Nuclear Power', *PennLive* (September 8, 2008), available online at www.pennlive.com/midstate/index.ssf/2008/09/midstate_supports_nuclear_powe.html (accessed March 17, 2016).
112 Quote from NEI's (Nuclear Energy Institute) Scott Peterson, from Duncan Mansfield, 'Nuclear Revival Begins with the Restart of TVA Reactor', *The Times* (May 5, 2007), available online at www.timesonline.com/nuclear-revival-begins-with-restart-of-tva-reactor/article_1695aa11-ae89-5dc3-ba0c-292c37a3dfa4.html (accessed July 27, 2016).
113 Tucker interview with Grace Halden.
114 Jacques Derrida, 'On a Newly Arisen Apocalyptic Tone in Philosophy', in *Raising the Tone of Philosophy: Late Essays by Immanuel Kant, Transformative Critique by Jacques Derrida*, ed. by Peter Fenves (London: The Johns Hopkins University Press, 1993), pp. 117–173 (p. 118).
115 David Herbert Lawrence, *Apocalypse* (Middlesex: Penguin, 1975), p. 3.
116 Lonna M. Malmsheimer, 'Three Mile Island: Fact, Frame, and Fiction', *American Quarterly*, 38.1 (1986), 35–52 (p. 40).
117 Malmsheimer, 'Three Mile Island', pp. 35–52 (p. 35).
118 Wade Allison, 'Viewpoint: We Should Stop Running Away from Radiation', *BBC* (March 26, 2011), available online at www.bbc.co.uk/news/world-12860842 (accessed May 15, 2016).
119 Emma M. Cappelluzzo, 'Nuclear Power and Educational Responsibilities', *The Phi Delta Kappan*, 61.1 (September 1979), 47–49.
120 See Cappelluzzo, 'Nuclear Power and Educational Responsibilities', pp. 47–49.
121 Peter Miller, 'Susquehanna: America's Small-Town River', *National Geographic*, 167.3 (March 1985) pp. 352–383 (p. 381).
122 Miller, 'Susquehanna: America's Small-Town River', p. 379.
123 See Brett McCollum, 'Nuclear Energy is a Safe Industry', in *Nuclear Power: A Reference Handbook*, 2nd edn, ed. by Henderson, pp. 130–137 (p. 130).

124 See Brett McCollum, 'Nuclear Energy is a Safe Industry', pp. 130–137 (p. 135).
125 See Brett McCollum, 'Nuclear Energy is a Safe Industry', pp. 130–137 (p. 135).
126 See Lloyd J. Dumas, 'Safety, Human Fallibility, and Nuclear Power', in *Nuclear Power: A Reference Handbook*, 2nd edn, ed. by Henderson, pp. 137–140 (p. 140).
127 'We All Live in Harrisburg', *The Economist*, April 7, 1979. Cited in *Three Mile Island Nuclear Powerplant Accident*, 'Hearings before the Subcommittee on Nuclear Regulation of the Committee on Environment and Public Works', United States Senate, Ninety-Sixth Congress, First Session (Washington D.C., April 10, 28, and 30, 1979), Part 1, p. 248.
128 Murray, *Nuclear Turnaround*, p. 53.
129 See Bradley C. Bower, 'TMI anniversary', *The Times* (March 28, 2011), available online at www.timesonline.com/tmi-anniversary/image_ec8fc588-5963-11e0-b6d7-0017a4a78c22.html (accessed July 13, 2016).
130 Quoted in Murray, *Nuclear Turnaround*, p. 53.

Documents

DOCUMENT 1

'This New Phenomenon'

In 1939, during World War II, Albert Einstein wrote to President F. D. Roosevelt about exciting developments in nuclear research. Initially, Einstein wished to convey information to the President about the exciting and beneficial potential for nuclear energy. However, as we can see in this letter, Einstein also drew Roosevelt's attention to the fact that nuclear bombs could also be constructed as a result of this research. Einstein also expressed concern over the fact that Germany was almost certainly working on the same fearsome technology. This letter displays the intimate connection nuclear energy has with nuclear warfare and how, in the early days of nuclear development, excitement as well as trepidation was experienced over the enormous potential of this innovation. Einstein's letter further expresses the determination for America to secure the new technology first; the fight for nuclear mastery would also come to dominate the Cold War. *See Chapter 2.*

ALBERT EINSTEIN IN A LETTER TO F. D. ROOSEVELT, AUGUST 2, 1939

Albert Einstein
Old Grove Road
Peconic, Long Island
August 2nd, 1939

F. D. Roosevelt
President of the United States
White House
Washington, D.C.

Sir:

Some recent work by E. Fermi and L. Szilard, which has been communicated to me in manuscript, leads me to expect that the element uranium may be turned into a new and important source of energy in the immediate future. Certain aspects of the situation which has arisen seem to call for watchfulness and if necessary, quick action on the part of the Administration. I believe therefore that it is my duty to bring to your attention the following facts and recommendations.

In the course of the last four months it has been made probable through the work of Joliot in France as well as Fermi and Szilard in America—that it may be possible to set up a nuclear chain reaction in a large mass of uranium, by which vast amounts of power and large quantities of new radium-like elements would be generated. Now it appears almost certain that this could be achieved in the immediate future.

This new phenomenon would also lead to the construction of bombs, and it is conceivable—though much less certain—that extremely powerful bombs of this type may thus be constructed. A single bomb of this type, carried by boat and exploded in a port, might very well destroy the whole port together with some of the surrounding territory. However, such bombs might very well prove too heavy for transportation by air.

The United States has only very poor ores of uranium in moderate quantities. There is some good ore in Canada and former Czechoslovakia, while the most important source of uranium is in the Belgian Congo.

In view of this situation you may think it desirable to have some permanent contact maintained between the Administration and the group of physicists working on chain reactions in America. One possible way of achieving this might be for you to entrust the task with a person who has your confidence and who could perhaps serve in an unofficial capacity. His task might comprise the following:

a) to approach Government Departments, keep them informed of the further development, and put forward recommendations for Government action, giving particular attention to the problem of securing a supply of uranium ore for the United States.

b) to speed up the experimental work, which is at present being carried on within the limits of the budgets of University laboratories, by providing funds, if such funds be required, through his contacts with private persons who are willing to make contributions for this cause, and perhaps also by obtaining co-operation of industrial laboratories which have necessary equipment.

I understand that Germany has actually stopped the sale of uranium from the Czechoslovakian mines which she has taken over. That she should have taken such early action might perhaps be understood on the ground that the son of the German Under-Secretary of State, von Weizsacker, is attached to the Kaiser-Wilhelm Institute in Berlin, where some of the American work on uranium is now being repeated.

Yours very truly,

A. Einstein

Albert Einstein

Credit: Albert Einstein, 'Letter to F. D. Roosevelt', 2 August 1939, in *AtomicArchive*, available online at www.atomicarchive.com/Docs/Begin/Einstein.shtml (accessed August 18, 2016).

DOCUMENT 2

Atoms for Peace

In Albert Einstein's letter to President F. D. Roosevelt in 1939, he conveyed that uranium could be used as a source of energy in the future. However, due to wartime pressures, attention fell to the application of this technology for the construction of nuclear weapons. America dropped the first nuclear bomb on Japan on August 6, 1945. In the 1950s there was a concentrated effort by President Dwight D. Eisenhower to utilize atomic energy for peace. The Atoms for Peace program, introduced in 1953, as well as the Atomic Energy Act of 1954, sought to shift the atom from the shadow of the mushroom cloud and reposition it for peaceful application. *See Chapter 2.*

PRESIDENT DWIGHT D. EISENHOWER TO THE 470TH PLENARY MEETING OF THE UNITED NATIONS GENERAL ASSEMBLY, DECEMBER 8, 1953. SPEECH EXCERPT

I feel impelled to speak today in a language that in a sense is new, one which I, who have spent so much of my life in the military profession, would have preferred never to use. That new language is the language of atomic warfare.

The atomic age has moved forward at such a pace that every citizen of the world should have some comprehension, at least in comparative terms, of the extent of this development, of the utmost significance to every one of us. Clearly, if the peoples of the world are to conduct an intelligent search for peace, they must be armed with the significant facts of today's existence.

My recital of atomic danger and power is necessarily stated in United States terms, for these are the only incontrovertible facts that I know, I need hardly point out to this Assembly, however, that this subject is global, not merely national in character.

On 16 July 1945, the United States set off the world's biggest atomic explosion. Since that date in 1945, the United States of America has conducted forty-two test explosions. Atomic bombs are more than twenty-five times as powerful as the weapons with which the atomic age dawned, while hydrogen weapons are in the ranges of millions of tons of TNT equivalent.

Today, the United States stockpile of atomic weapons, which, of course, increases daily, exceeds by many times the total equivalent of the total of all bombs and all shells that came from every plane and every gun in every theatre of war in all the years of the Second World War. A single air group whether afloat or land based, can now deliver to any reachable target a destructive cargo exceeding in power all the bombs that fell on Britain in all the Second World War.

In size and variety, the development of atomic weapons has been no less remarkable. The development has been such that atomic weapons have virtually achieved conventional status within our armed services. In the United States, the Army, the Navy, the Air Force and the Marine Corps are all capable of putting this weapon to military use.

But the dread secret and the fearful engines of atomic might are not ours alone.

In the first place, the secret is possessed by our friends and allies, the United Kingdom and Canada, whose scientific genius made a tremendous contribution to our original discoveries and the designs of atomic bombs.

The secret is also known by the Soviet Union. The Soviet Union has informed us that, over recent years, it has devoted extensive resources to atomic weapons. During this period the Soviet Union has exploded a series of atomic devices, including at least one involving thermo-nuclear reactions.

If at one time the United States possessed what might have been called a monopoly of atomic power, that monopoly ceased to exist several years ago. Therefore, although our earlier start has permitted us to accumulate what is today a great quantitative advantage, the atomic realities of today comprehend two facts of even greater significance. First, the knowledge now possessed by several nations will eventually be shared by others, possibly all others.

Second, even a vast superiority in numbers of weapons, and a consequent capability of devastating retaliation, is no preventive, of itself, against the fearful material damage and toll of human lives that would be inflicted by surprise aggression.

The free world, at least dimly aware of these facts, has naturally embarked on a large program of warning and defense systems. That program will be accelerated and extended. But let no one think that the expenditure of vast sums for weapons and systems of defense can guarantee absolute safety for the cities and citizens of any nation. The awful arithmetic of the atomic bomb doesn't permit of any such easy solution. Even against the most powerful defense, an aggressor in possession of the effective minimum number of atomic bombs for a surprise attack could probably place a sufficient number of his bombs on the chosen targets to cause hideous damage.

Should such an atomic attack be launched against the United States, our reactions would be swift and resolute. But for me to say that the defense capabilities of the United States are such that they could inflict terrible losses upon an aggressor, for me to say that the retaliation capabilities of the United States are so great that such an aggressor's land would be laid waste, all this, while fact, is not the true expression of the purpose and the hopes of the United States.

To pause there would be to confirm the hopeless finality of a belief that two atomic colossi are doomed malevolently to eye each other indefinitely across a trembling world. To stop there would be to accept helplessly the probability of civilization destroyed, the annihilation of the irreplaceable heritage of mankind handed down to us from generation to generation, and the condemnation of mankind to begin all over again the age-old struggle upward from savagery towards decency, and right, and justice. Surely no sane member of the human race could discover victory in such desolation. Could anyone wish his name to be coupled by history with such human degradation and destruction? Occasional pages of history do record the faces of the "great destroyers", but the whole book of history reveals mankind's never-ending quest for peace and mankind's God-given capacity to build.

It is with the book of history, and not with isolated pages, that the United States will ever wish to be identified. My country wants to be constructive, not destructive. It wants agreements, not wars, among nations. It wants itself to live in freedom and in the confidence that the peoples of every other nation enjoy equally the right of choosing their own way of life.

So my country's purpose is to help us to move out of the dark chamber of horrors into the light, to find a way by which the minds of men, the hopes of men, the souls of men everywhere, can move forward towards peace and happiness and well-being.

In this quest, I know that we must not lack patience. I know that in a world divided, such as ours today, salvation cannot be attained by one dramatic act. I know that many steps will have to be taken over many months before the world can look at itself one day and truly realize that a new climate of mutually peaceful confidence is abroad in the world. But I know, above all else, that we must start to take these steps—now.

The United States and its allies, the United Kingdom and France, have over the past months tried to take some of these steps. Let no one say that we shun the conference table. On the record has long stood the request of the United States, the United Kingdom and France to negotiate with the Soviet Union the problems of a divided Germany. On that record has long stood the request of the same three nations to negotiate an Austrian peace treaty. On the same record still stands the request of the United Nations to negotiate the problems of Korea.

Most recently we have received from the Soviet Union what is in effect an expression of willingness to hold a four-Power meeting. Along with our allies, the United Kingdom and France, we were pleased to see that this note did not contain the unacceptable pre-conditions previously put forward. As you already know from our joint Bermuda communique, the United States, the United Kingdom and France have agreed promptly to meet with the Soviet Union.

The Government of the United States approaches this conference with hopeful sincerity. We will bend every effort of our minds to the single purpose of emerging from that conference with tangible results towards peace, the only true way of lessening international tension.

We never have, and never will, propose or suggest that the Soviet Union surrender what rightly belongs to it. We will never say that the peoples of the USSR are an enemy with whom we have no desire ever to deal or mingle in friendly and fruitful relationship.

On the contrary, we hope that this coming conference may initiate a relationship with the Soviet Union which will eventually bring about a freer mingling of the peoples of the East and of the West—the one sure, human way of developing the understanding required for confident and peaceful relations.

Instead of the discontent which is now settling upon Eastern Germany, occupied Austria and the countries of Eastern Europe, we seek a harmonious family of free European nations, with none a threat to the other, and least of all a threat to the peoples of the USSR. Beyond the turmoil and strife and misery of Asia, we seek peaceful opportunity for these peoples to develop their natural resources and to elevate their lot.

These are not idle words or shallow visions. Behind them lies a story of nations lately come to independence, not as a result of war, but through free grant or peaceful negotiation. There is a record already written of assistance gladly given by nations of the West to needy peoples and to those suffering the temporary effects of famine, drought and natural disaster. These are deeds of peace. They speak more loudly than promises or protestations of peaceful intent.

But I do not wish to rest either upon the reiteration of past proposals or the restatement of past deeds. The gravity of the time is such that every new avenue of peace, no matter how dimly discernible, should be explored.

There is at least one new avenue of peace which has not been well explored—an avenue now laid out by the General Assembly of the United Nations.

In its resolution of 28 November 1953 (resolution 715 (VIII)) this General Assembly suggested: "that the Disarmament Commission study the desirability of establishing a sub-committee consisting of representatives of the Powers principally involved, which should seek in private an acceptable solution and report . . . on such a solution to the General Assembly and to the Security Council not later than 1 September 1954."

The United States, heeding the suggestion of the General Assembly of the United Nations, is instantly prepared to meet privately with such other countries as may be "principally involved", to seek "an acceptable solution" to the atomic armaments race which overshadows not only the peace, but the very life, of the world.

We shall carry into these private or diplomatic talks a new conception. The United States would seek more than the mere reduction or elimination of atomic materials for military purposes. It is not enough to take this weapon out of the hands of the soldiers. It must be put into the hands of those who will know how to strip its military casing and adapt it to the arts of peace.

The United States knows that if the fearful trend of atomic military build-up can be reversed, this greatest of destructive forces can be developed into a great boon, for the benefit of all mankind. The United States knows that peaceful power from atomic energy is no dream of the future. The capability, already proved, is here today. Who can doubt that, if the entire body of the world's scientists and engineers had adequate amounts of fissionable material with which to test and develop their ideas, this capability would rapidly be transformed into universal, efficient and economic usage?

To hasten the day when fear of the atom will begin to disappear from the minds the people and the governments of the East and West, there are certain steps that can be taken now.

I therefore make the following proposal.

The governments principally involved, to the extent permitted by elementary prudence, should begin now and continue to make joint contributions from their stockpiles of normal uranium and fissionable materials to an international atomic energy agency. We would expect that such an agency would be set up under the aegis of the United Nations. The ratios of contributions, the procedures and other details would properly be within the scope of the "private conversations" I referred to earlier.

The United States is prepared to undertake these explorations in good faith. Any partner of the United States acting in the same good faith will find the United States a not unreasonable or ungenerous associate.

Undoubtedly, initial and early contributions to this plan would be small in quantity. However, the proposal has the great virtue that it can be undertaken without the irritations and mutual suspicions incident to any attempt to set up a completely acceptable system of world-wide inspection and control.

The atomic energy agency could be made responsible for the impounding, storage and protection of the contributed fissionable and other materials. The ingenuity of our scientists will provide special safe conditions under which such a bank of fissionable material can be made essentially immune to surprise seizure.

The more important responsibility of this atomic energy agency would be to devise methods whereby this fissionable material would be allocated to serve the peaceful pursuits of mankind. Experts would be mobilized to apply atomic energy to the needs of agriculture, medicine and other peaceful activities. A special purpose would be to provide abundant electrical energy in the power-starved areas of the world.

Thus the contributing Powers would be dedicating some of their strength to serve the needs rather than the fears of mankind.

The United States would be more than willing—it would be proud to take up with others "principally involved" the development of plans whereby such peaceful use of atomic energy would be expedited.

Of those "principally involved" the Soviet Union must, of course, be one.

I would be prepared to submit to the Congress of the United States, and with every expectation of approval, any such plan that would, first, encourage world-wide investigation into the most effective peacetime uses of fissionable material, and with the certainty that the investigators had all the material needed for the conducting of all experiments that were appropriate; second, begin to diminish the potential destructive power of the world's atomic stockpiles; third, allow all peoples of all nations to see that, in this enlightened age, the great Powers of the earth, both of the East and of the West, are interested in human aspirations first rather than in building up the armaments of war; fourth, open up a new channel for peaceful discussion and initiative at least a new approach to the many difficult problems that must be solved in both private and public conversations if the world is to shake off the inertia imposed by fear and is to make positive progress towards peace.

Against the dark background of the atomic bomb, the United States does not wish merely to present strength, but also the desire and the hope for peace. The coming months will be fraught with fateful decisions. In this Assembly, in the capitals and military headquarters of the world, in the hearts of men everywhere, be they governed or governors, may they be the decisions which will lead this world out of fear and into peace.

To the making of these fateful decisions, the United States pledges before you, and therefore before the world, its determination to help solve the fearful atomic dilemma—to devote its entire heart and mind to finding the way by which the miraculous inventiveness of man shall not be dedicated to his death, but consecrated to his life.

Figure D.1 The Atoms for Peace symbol

United States Atomic Energy Commission (1955)

The Atoms for Peace symbol was used as part of the Geneva International Conference on the Peaceful Uses of Atomic Energy *Conference (1955). Four icons decorate the image; the symbols of agriculture (left), science (top), medicine (right), and industry (bottom). These symbols of modern life surround the atom, which is central in both the image and, it is implied, in modern activity. To the left and right of the image, in the outer circle, two olive branches represent the safe and peaceful use of the atom—an olive branch, essentially, to the world.*

Credits: Dwight D. Eisenhower, 'Atoms for Peace', 470th Plenary Meeting of the United Nations General Assembly, 8 December 1953, in *IAEA*, available online at www.iaea.org/about/history/atoms-for-peace-speech (accessed July 14, 2016).

DOCUMENT 3

'An Unprecedented Accident'

The announcement in 1967 that Pennsylvania's Susquehanna River would gain a new power plant did not meet with much press attention. Nuclear power was very much seen, by many, to be the peaceful application that Eisenhower promised all those years ago in his 'Atoms for Peace' speech. When the Three Mile Island nuclear plant featured in snippet news stories, focus mainly fell on the successes of the plant and its promise of jobs and opportunities. However, in March 1979, the Three Mile Island plant would be promoted from minor articles in local newspapers and would feature in major international publications as America's worst commercial nuclear disaster.

The crisis occurred at a time in which nuclear interest was in decline and ambivalence was high. The shock of this unprecedented crisis shattered prevailing ideas that nuclear power was the completely safe and innocent twin of the nuclear weapon; suddenly the media was expressing anxiety about radiation releases and the potential for an explosion. The crisis was so traumatic and such a landmark event that President Jimmy Carter visited the plant to ascertain for himself the extent of the crisis and to calm escalating fears. Following the crisis numerous reports ranging from psychological studies, to health reports, to official investigations, were published. The Kemeny Report (The President's Commission on the Accident at Three Mile Island, created by President Carter and led by John Kemeny, 1979) and the Rogovin Report (a Nuclear Regulatory Commission investigation led by Mitchell Rogovin, 1980) are leading documents that explore the events leading up to the crisis and detail essential improvements required for safer operation. An excerpt here, from the epilogue of the Kemeny Report, gives a brief overview of some ways the local community was affected by the events of March and April 1979. *See Chapters 3 and 5.*

Figure D.2 President Jimmy Carter leaving Three Mile Island

President's Commission on the Accident at Three Mile Island, March 29–April 30, 1979

'EPILOGUE' FROM THE 'KEMENY REPORT' (REPORT OF THE PRESIDENT'S COMMISSION ON THE ACCIDENT AT THREE MILE ISLAND)

The accident at Three Mile Island did not end with the breaking up of the bubble, nor did the threat to the health and safety of the workers and the community suddenly disappear. A small bubble remained, gases still existed within TMI-2's cooling water, and the reactor itself was badly damaged. Periodic releases of low-level radiation continued, and some feared a major release of radioactive iodine-131 might yet occur. Schools remained closed. The Governor's recommendation that pregnant women and preschool children stay more than 5 miles from the plant continued.

Saturday, March 31, the Department of Health, Education, and Welfare had arranged for the rapid manufacture of nearly a quarter million bottles of potassium iodide.[1] That same day, the Pennsylvania Bureau of Radiation Protection—which had originally accepted HEW's offer to obtain the drug—transferred responsibility for handling the radioactive iodine blocker to the state's Department of Health. Gordon MacLeod, who headed the health department, put the drug shipments in a warehouse as they began arriving Sunday. During the weekend, Thomas Gerusky, director of the

Bureau of Radiation Protection, requested that his people at TMI be issued potassium iodide; Gerusky wanted BRP personnel to have the thyroid-blocking agent available should a release of radioactive iodine occur. MacLeod refused. He argued that if the public learned that any of the drug had been issued, a demand for its public distribution would result.

MacLeod had the backing of the Governor's office and Harold Denton in his decision not to issue the potassium iodide. The decision did not find agreement in Washington, however. On Monday, Jack Watson asked HEW to prepare recommendations for the drug's distribution and use. These were developed by a group headed by Donald Frederickson, director of the National Institutes of Health. The recommendation included: administering potassium iodide immediately to all workers on the Island; providing the drug to all people who would have less than 30 minutes' warning of a radioactive iodine release (roughly those within 10 miles of the plant); and that local authorities assess these recommendations in light of their first-hand knowledge of the situation.

Governor Thornburgh received the recommendations in a White House letter on Tuesday, although some Pennsylvania officials had learned of them Monday. MacLeod strongly opposed distributing the drug to the public. Among his reasons: radioiodine levels were far below what was indicated for protective action, and the likelihood of a high-level release from TMI-2 was diminishing; distributing the drug would increase public anxiety and people might take it without being told to do so; and the possibility of adverse side-effects presented a potential public health problem in itself. MacLeod chose not to accept the federal recommendations. The potassium iodide remained in a warehouse under armed guard throughout the emergency. In midsummer, the FDA moved the drug to Little Rock, Arkansas, for storage.

★ ★ ★

Tuesday, April 3, General Public Utilities, Met Ed's parent company, established its TMI-2 recovery organization to oversee and direct the long process of cleaning up TMI-2. Robert Arnold, a vice president of another subsidiary, the GPU Service Corporation, was named to head the recovery operation.[2]

Wednesday, April 4, schools outside the 5-mile area surrounding TMI reopened. All curfews were lifted. But schools within 5 miles of the Island remained closed and the Governor's advisory remained in effect for pregnant women and preschool children.

Some sense of normalcy was gradually returning to the TMI area. Governor Thornburgh asked Denton repeatedly if the advisory could be lifted, allowing pregnant women and preschool children to return home.

But the NRC wanted some specific event as a symbol to announce the crisis had ended. At first, the NRC looked to reaching "cold shutdown"—the point at which the temperature of TMI-2's reactor coolant fell below the boiling point of water. When it became obvious that cold shutdown was days away, agreement was reached between Pennsylvania's Bureau of Radiation Protection and the NRC on ending the advisory. On Saturday, April 7, Kevin Molloy, at the request of the Governor's office, read a press release announcing the closing of the evacuation shelter at the Hershey Park Arena. Not until 2 days later, however, did Governor Thornburgh officially withdraw the advisory.[3]

* * *

The accident at TMI did not end with cold shutdown, nor will it end for some time. More than a million gallons of radioactive water remain inside the containment building or stored in auxiliary building tanks. The containment building also holds radioactive gases and the badly damaged and highly radioactive reactor core. Radioactive elements contaminate the walls, floors, and equipment of several buildings. Ahead lies a decontamination effort unprecedented in the history of the nation's nuclear power industry—a cleanup whose total cost is estimated at $80 to $200 million and which will take several years to complete.[4]

NOTES

1 Original footnote: 122. For a complete recount of the potassium iodide story, see "Report of the Public Health and Epidemiology Task Group" and "Report of the Office of Chief Counsel on Emergency Response."
2 Original Footnote: 123. For discussion of Met Ed's recovery efforts, see section on TMI-2 recovery program in "Report of the Office of Chief Counsel on the Role of the Managing Utility and Its Suppliers" and technical staff analysis report on "Recovery."
3 Original Footnote: 124. Molloy deposition, pp. 115–117. See also "Report of the Office of Chief Counsel on Emergency Response" and "Technical Staff Report on Emergency Preparedness and Response."
4 Original Footnote: 125. For a discussion of TMI-2's recovery program, see technical staff analysis report on "Recovery" and the TMI-2 recovery program section of "Report of the Office of Chief Counsel on the Role of the Managing Utility and Its suppliers."

Credit: John G. Kemeny, *et al.*, *The Need for Change: The Legacy of TMI*, 'Report of the President's Commission on the Accident at Three Mile Island' (Washington D.C., October 1979), pp. 137–138.

DOCUMENT 4

Nuclear Imaginings

How Popular Culture Responded

Writing in 1980, Walter C. Patterson commented:

> Even now, well before the full history of the TMI-2 accident can be written, it has already passed into folklore. The very name "Three Mile Island" now conveys a message, not only in English but also in many other languages. It is not easy, however, to be precise about the substance of the message.

Popular culture—art, music, literature, television, and film—are vital mediums through which to explore various messages emanating from crises experienced by the public. Here is one example [see Figure D.3].

The Dispatch *became an interesting companion to California's Savanna High School yearbooks as, according to James A. Ollinger, the newspaper acted as a snapshot of interests and events on a daily basis. One such interest was nuclear power. In this comical cartoon from* The Dispatch *in 1982, artist Dan Kelton reflects on the public anxieties that nuclear plants pose a threat to human health. During Three Mile Island, pregnant women were advised to evacuate and many were concerned about radiation endangering infants; this cartoon reflects on these fears. The accompanying article by Kelton makes reference to the accident at Three Mile Island, while also commenting that, despite perceived risks, nuclear power is an important and inescapable part of American energy generation. Kelton's cartoon draws attention not only to the endurance of Three Mile Island in popular culture, but also highlights the relevance of the Pennsylvanian event years later and in another state. See Chapter 4.*

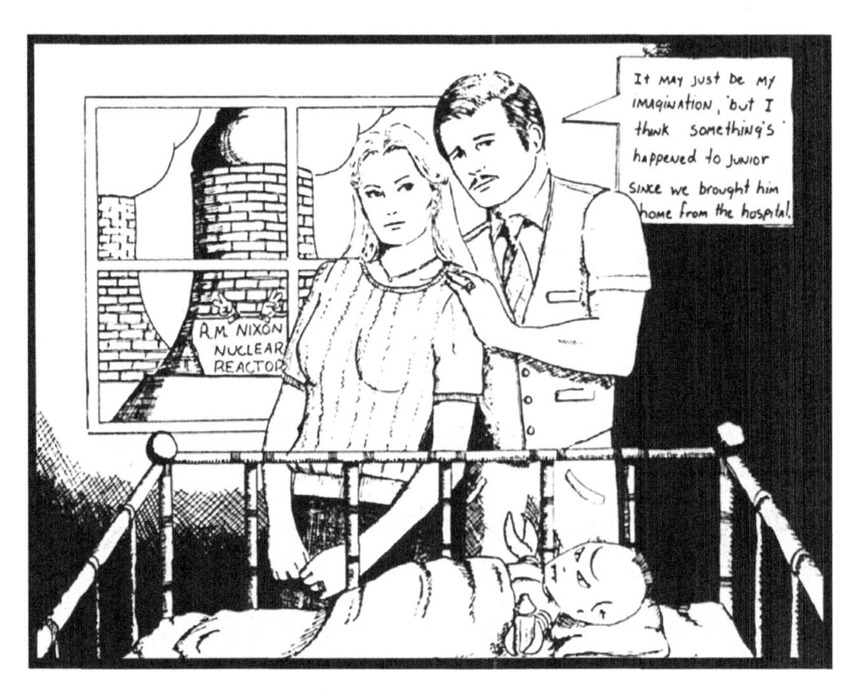

Figure D.3 Cartoon by Dan Kelton from *The Dispatch*

Dan Kelton, *The Dispatch*, January 22, 1982, p. 2

Credits: Walter C. Patterson, in *Impact Abroad of the Accident at the Three Mile Island Nuclear Power Plant*, 'Subcommittee on Energy, Nuclear Proliferation and Federal Services of the Committee on Governmental Affairs', United States Senate (Washington: Library of Congress, May 1980), p. 21.

Bibliography

Academic Text: Books, Articles, and Journals

Agar, Jon, 'What happened in the sixties?', *The British Journal for the History of Science* 41.4 (2008), pp. 567–600

The American Atom. A Documentary History of Nuclear Policies from the Discovery of Fission to the Present 1939–1984, ed. by Robert C. Williams and Philip L. Cantelon (Philadelphia: University of Pennsylvania Press, 1984)

Amerisov, Alexander, 'Chernobyl Emerging Story: A Chronology of Soviet Media Coverage', *Bulletin of the Atomic Scientists*, 3.1 (September 1986), pp. 38–39

Arendt, Hannah, *The Origins of Totalitarianism* (London: Harvest, 1976)

—— *The Human Condition*, trans. by Margaret Canovan, 2nd edn (London: University of Chicago Press, 1998)

Arthur, W. Brian, *The Nature of Technology: What It Is and How It Evolves* (London: Penguin, 2010)

The Atomic Age Opens, ed. by Donald Porter Geddes and Gerald Wendt (New York: World Publishing Company, 1945)

Baudrillard, Jean, 'The China Syndrome', in *Simulacra and Simulations*, by Jean Baudrillard, trans. by Shelia Faria Glaser (Michigan: University of Michigan Press, 2012), pp. 53–58

Beaver, William, 'The Failed Promise of Nuclear Power', *Independent Review* 15.3 (2011), pp. 399–411

Bess, Michael, *Realism, Utopia, and the Mushroom Cloud. Four Activist Intellectuals and Their Strategies for Peace, 1945–1989* (London: University of Chicago Press, 1993), pp. 60–61

Bethe, Hans, 'Hans Bethe, Comments on the History of the H-Bomb', in *The American Atom. A Documentary History of Nuclear Policies from the Discovery of Fission to the Present. 1939–1984*, ed. by Robert Williams and Philip Cantelon (Philadelphia: University of Pennsylvania Press, 1984), pp. 132–141

Boyer, Paul, *By the Bomb's Early Light: American Thought and Culture at the Atomic Age* (New York: The University of North Carolina Press, 1994)

—— 'From Activism to Apathy: America and the Nuclear Issue, 1963–1980, *Bulletin of the Atomic Scientist*, 40.7 (August–September 1984), pp. 14–15

Brians, Paul, 'Nuclear War Fiction for Young Readers: A Commentary and Annotated Bibliography', in *Science Fiction, Social Conflict and War*, ed. by John Philip Davies (Manchester: Manchester University Press, 1990), pp. 132–151

Broderick, Mick, 'Releasing the Hounds: The Simpsons as Anti-Nuclear Satire, in *Leaving Springfield*, ed. by John Alberti (Detriot: Wayne State University Press, 2004), pp. 244–272

Brombert, Victor, *Musings on Mortality: From Tolstoy to Primo Levi* (London: The University of Chicago Press, 2013)

Bromet, Evelyn J., David K. Parkinson and Leslie O. Dunn, 'Long-term Mental Health Consequences of the Accident at Three Mile Island', *International Journal of Mental Health*, 19.2 (Summer 1990), pp. 48–60

Brosnan, John, *Future Tense. The Cinema of Science Fiction* (London: Macdonald and Jane's Publishers, 1978)

Brubaker, Jack, *Down the Susquehanna to the Chesapeake* (Pennsylvania: The Pennsylvania State University Press, 2002)

Brzezinski, Matthew, *Red Moon Rising: Sputnik and the Rivalries that Ignited the Space Race* (London: Bloomsbury, 2008)

Bukatman, Scott, 'There's Always Tomorrowland: Disney and the Hypercinematic Experience', *Reviewed Works*, 57 (1991), pp. 55–78

Bulletin of the Atomic Scientists, 'Background and Mission: 1945–2016', *thebulletin.org* (2016), available online at http://thebulletin.org/background-and-mission-1945–2016 (accessed August 8, 2016)

—— 'Timeline', *thebulletin.org* (2016), available online at http://thebulletin.org/timeline (accessed August 8, 2016)

Bunn, George and Fritz Steinhausler, 'Guarding Nuclear Reactors and Material from Terrorists and Thieves', *Arms Control Today*, 31.8 (October 2001), pp. 8–12

Burleson, Clyde W., *The Day the Bomb Fell: True Stories of the Nuclear Age* (London: Sphere, 1980)

Cable, Sherry, Edward J. Walsh and Rex H. Warland, 'Differential Paths to Political Activism: Comparisons of Four Mobilization Processes after the Three Mile Island Accident', *Social Forces*, 66.4 (June 1988), pp. 951–969

Cappelluzzo, Emma M., 'Nuclear Power and Educational Responsibilities', *The Phi Delta Kappan*, 61.1 (September 1979), pp. 47–49

Carter, Luther J., 'The "Movement" Moves on to Antinuclear Protest', *Science*, 204.4394 (May 18, 1979)

Chisholm, Rupert F. and Stanislav V. Kasl, 'The Effects of Work Site, Supervisory Status, and Job Function on Nuclear Workers' Responses to the TMI Accident', *Journal of Occupational Behavior*, 3.1 (January, 1982), pp. 39–62

The Committee for the Complication of Materials of Damage Caused by the Atomic Bomb in Hiroshima and Nagasaki, *Hiroshima and Nagasaki: The Physical, Medical, Social Effects of the Atomic Bombings*, trans. by Eisei Ishikawa and David L. Swain (London: Hutchinson, 1981)

'Conference: Reporting Scientific and Technological Controversy', *Science, Technology, & Human Values*, 5.32 (Summer, 1980), pp. 34–35

Cottrell, Alan, *How Safe is Nuclear Energy* (London: Heinemann, 1981)

Croall, Stephen and Kaianders Sempler, *Nuclear Power for Beginners* (London: Beginners Books, 1978)

Dass-Brailsford, Priscilla, *A Practical Approach to Trauma: Empowering Interventions* (London: Sage, 2007)

Davies, Sarah, *Popular Opinion in Stalin's Russia: Terror, Propaganda, and Dissent, 1934–1941* (Cambridge: Cambridge University Press, 1997)

Del Tredici, Robert, *The People of Three Mile Island* (San Francisco: Sierra Club Books, 1980)

Denning, Michael, *The Cultural Front: The Laboring of American Culture in the Twentieth Century* (London: Verso, 1997)

Derkins, Susie, *When Disaster Strikes: The Meltdown at Three Mile Island* (New York: The Rosen Publishing Group, 2003)

Derrida, Jacques, 'On a Newly Arisen Apocalyptic Tone in Philosophy', in *Raising the Tone of Philosophy: Late Essays by Immanuel Kant, Transformative Critique by Jacques Derrida*, ed. by Peter Fenves (London: The John Hopkins University Press, 1993), pp. 117–173

—— *Spectres of Marx: The State of Debt, the Work of Mourning and the New International*, trans. by Peggy Kamuf (New York: Routledge, 1994)

Dew, M. A. and E. J. Bromet, 'Predictors of Temporal Patterns of Psychiatric Distress During 10 Years Following the Nuclear Accident at Three Mile Island', *Social Psychiatry and Psychiatric Epidemiology*, 28 (1993), pp. 49–55

Dumas, Lloyd J., 'Safety, Human Fallibility, and Nuclear Power', in *Nuclear Power: A Reference Handbook*, 2nd edn, ed. by Henderson, pp. 137–140

'Earthrise', NASA (2015) available online at www.nasa.gov/multimedia/image gallery/image_feature_1249.html (accessed April 13, 2016)

Eggert, Gerald G., *Harrisburg Industrializes: The Coming of Factories to an American Community* (Pennsylvania: Pennsylvania University Press, 1993)

Ferguson, Charles D., 'Japan Melted Down, but that Doesn't Mean the End of the Atomic Age', *Foreign Policy*, 189 (November 2011), pp. 49–53

Flack, J., 'Scapegoat for Stress', *Science News*, 127.18 (May 4, 1985), p. 286

Florman, Samuel C., *The Existential Pleasures of Engineering* (New York: St. Martin's Press, 1976)

Ford, Daniel F., *Three Mile Island: Thirty Minutes to Meltdown* (New York: Viking Press, 1982)

Francis, John and Paul Abrecht, 'Foreword', in *Facing up to Nuclear Power*, ed. by John Francis and Paul Abrecht (Philadelphia: The Westminster Press, 1976), pp. 1–4

Freudenburg, William R. and Timothy R. Jones, 'Attitudes and Stress in the Presence of Technological Risk: A Test of the Supreme Court Hypothesis', *Social Forces*, 69.4 (June, 1991), pp. 1143–1168

Friedman, Sharon M., 'Three Mile Island, Chernobyl, and Fukushima: An Analysis of Traditional and New Media Coverage of Nuclear Accidents and Radiation', *Bulletin of the Atomic Scientists*, 67.5 (September 2011), pp. 55–65

Fuller, Buckminster, 'Introduction', in *Three Mile Island: Turning Point*, by Bill Keisling (Washington: Express Publications, 1980)

Fuller, John, *We Almost Lost Detroit* (New York: Ballantine Books, 1976)

Gamson, William A. and Andre Modigliani, 'Media Discourse and Public Opinion on Nuclear Power: A Constructionist Approach', *American Journal of Sociology*, 95.1 (1989), pp. 1–37

Garcia-Vasquez, Marina, 'Nuclear Waste Is Art in the Work of Taryn Simon', *The Creators Project* (June 9, 2015), available online at http://thecreators project.vice.com/en_uk/blog/nuclear-waste-is-art-in-the-work-of-taryn-simon (accessed May 17, 2016)

Garrett, Benjamin C. and John Hart, *Historical Dictionary of Nuclear, Biological and Chemical Warfare* (Plymouth: Scarecrow Press, 2007)

Garrison, Jim, *The Plutonium Culture: From Hiroshima to Harrisburg* (New York: Continuum, 1981)

Gerusky, Thomas M., 'Three Mile Island: Assessment of Radiation Exposures and Environmental Contamination', in *The Three Mile Island Nuclear Accident: Lessons and Implications*, ed. by Thomas H. Moss and David L. Sills (New York: The New York Academy of Sciences, 1981), pp. 54–62

Ginzburg, Harold M., 'The Psychological Consequences of the Chernobyl Accident: Findings from the International Atomic Energy Agency Study', *Public Health Reports*, 108.2 (March–April, 1993), pp. 184–192

Goldhaber, Marilyn K., *et al.*, 'The Three Mile Island Population Registry', *Public Health Reports*, 98.6 (November–December, 1983), pp. 603–609

Goldsteen, Raymond L., Karen Goldsteen and John K. Schorr, 'Trust and Its Relationship to Psychological Distress: The Case of Three Mile Island', *Political Psychology*, 13.4 (December, 1992), pp. 693–707

Goldstine, Herman H., *The Computer: From Pascal to von Neumann* (West Sussex: Princeton University Press, 1993)

Gundersen, Arnie, 'Fukushima', *Fairewinds* (n.d.), available online at www.fairewinds.org/fukushima (accessed July 20, 2016)

—— 'Nuclear Science Guy Postulates Meltdown Every Seven Years', *Huntington News.net* (February 9, 2015), available online at www.huntingtonnews.net/106198 (accessed March 18, 2016)

Gundersen, Maggie, 'TMI: A Human Perspective', *Fairewinds* (April 2, 2015), available online at www.fairewinds.org/demystify/three-mile-island-opera?rq=Three%20mile%20island%20opera (accessed August 2, 2016)

Gyorgy, Anna, *No Nukes: Everyone's Guide to Nuclear Power* (Montreal: Black Rose Books, 1979)

Halevy, Libbe, *Yes, I Glow in the Dark! One Mile from Three Mile Island to Fukushima and Beyond* ((n.p.): Heartistry Communications, 2014), Kindle ebook

Hatch, Maureen C., *et al.*, 'Cancer Near the Three Mile Island Nuclear Plant: Radiation Emissions', *Journal of Epidemiology*, 132.3 (September 1990), pp. 397–412

Heidegger, Martin, *The Question Concerning Technology and Other Essays*, trans. by William Lovitt (London: Harper and Row, 1977)

Henderson, Harry, *Nuclear Power: A Reference Handbook*, 2nd edn (California: ABC-CLIO, 2014)

Hesse-Honegger, Cornelia, 'Field studies in radiation-contaminated areas in the U.S.', (n.d.), available online at www.wissenskunst.ch/uk/completed-studies/united-states/1/ (accessed May 18, 2016)

Hilfer, Tony, *American Fiction Since 1940* (Harlow: Longman, 1992)

Hogan, Michael J., 'Hiroshima in History and Memory: An Introduction', in *Hiroshima in History and Memory* (Cambridge: Cambridge University Press, 1999), pp. 1–10

Houts, Peter S., Paul D. Cleary and Teh-Wei Hu, *Three Mile Island Crisis. Psychological, Social, and Economical Impacts on the Surrounding Population* (London: Pennsylvania State University Press, 1988)

Jacks, G. V. and R. O. White, *The Rape of the Earth: A World Survey of Soil Erosion* (London: Faber and Faber, sixth impression 1949)

Jacobs, Renée, *Slow Burn: A Photodocument of Centralia, Pennsylvania* (Pennsylvania: Pennsylvania University Press, 2010)

Johnson, James H., Jr. and Donald J. Zeigler, 'Distinguishing Human Responses to Radiological Emergencies', *Economic Geography*, 59.4 (October, 1983), pp. 386–402

Kafka, Janet, 'Why Science Fiction?', *The English Journal*, 64.5 (May, 1975), 46–53

Kharecha, Pushker A. and James E. Hansen, 'Prevented Mortality and Greenhouse Gas Emissions from Historical and Projected Nuclear Power', *Environmental, Science & Technology*, 47.9 (2013), pp. 4889–4895

Kirsch, Scott, *Proving Grounds: Project Plowshare and the Unrealized Dream of Nuclear Earthmoving* (New Jersey: Rutgers University Press, 2005)

Klaidman, Stephen, *Health in the Headlines* (New York: Oxford University Press, 1991)

Kline, Stephan J., 'What is Technology', *Bulletin of Science, Technology & Society*, 1.215 (1985), pp. 215–218

Koppes, Clayton R. and Gregory D. Black, *Hollywood Goes to War. How Politics, Profits and Propaganda Shaped World War Two Movies* (Los Angeles: University of California Press, 1990)

Lamarsh, John R., 'Safety Considerations in the Design and Operation of Light Water Nuclear Power Plants', in *The Three Mile Island Nuclear Accident: Lessons and Implications*, ed. by Thomas H. Moss and David L. Sills (New York: The New York Academy of Sciences, 1981), pp. 13–19

Langer, Lawrence L., *Holocaust Testimonies: The Ruins of Memory* (London: Yale University Press, 1991)

Lanouette, William, *Genius in the Shadows. A Biography of Leo Szilard, the Man Behind the Bomb* (Chicago: The University of Chicago Press, 1994)

Lavine, Matthew, *The First Atomic Age. Scientists, Radiations, and the American Public 1895–1945* (New York: Palgrave Macmillan, 2013)

Lawrence, David Herbert, *Apocalypse* (Middlesex: Penguin, 1975)

Levi, Primo, *The Drowned and the Saved*, trans. by Raymond Rosenthal (London: Joseph, 1986)

Lewis, Michael L., 'From Science to Science Fiction: Leo Szilard and Fictional Persuasion', in *The Writing on the Cloud: American Culture Confronts the Atomic Bomb*, ed. by Alison M. Scott and Christopher D. Geist (Oxford: University Press of America, 1997), pp. 95–105

Lief, Michael S., H. Mitchell Caldwell and Benjamin Bycel, *Ladies and Gentlemen of the Jury: Greatest Closing Arguments in Modern Law* (New York: Scribner, 1998)

Lifton, Robert J., *The Nazi Doctors. Medical Killing and the Psychology of Genocide* (New York: Basic Books, 2000)

—— 'NUCLEAR', *New York Times*, September 26, 1982

—— 'On the Nuclear Altar', *New York Times*, July 26, 1979

Lifton, Robert J. and Richard Falk, *Indefensible Weapons: The Political and Psychological Case Against Nuclearism* (New York: Basic Books, 1982)

Lifton, Robert J. and Eric Markusen, *The Genocide Mentality* (New York: Basic Books, 1990)

Lifton, Robert J. and Greg Mitchell, *Hiroshima in America, Fifty Years of Denial* (New York: A Grosset/Putnam Book, 1995)

Mahaffey, James, *Atomic Awakening* (New York: Pegasus Books, 2009

—— *Nuclear Power: This History of Nuclear Power* (New York: Facts on File, 2011)

Malmsheimer, Lonna M., 'Three Mile Island: Fact, Frame, and Fiction', *American Quarterly*, 38.1 (1986), pp. 35–52

Marc, David, *Comic Visions: Television Comedy and American Culture*, 2nd edn (Massachusetts: Blackwell, 1997)

Martin, Richard, 'Detonating on Canvas: The Abstract Bomb in American Art', in *The Writing on the Cloud: American Culture Confronts the Atomic Bomb*, ed. by Alison M. Scott and Christopher D. Geist (Oxford: University Press of America, 1997), pp. 73–79

Marx, Leo, *The Machine in the Garden: Technology and the Pastoral Ideal in America* (Oxford: Oxford University Press, 2000)

May, Timothy, 'The End of Enthusiasm: Science and Technology', in *The Columbia Guide to America in the 1960s*, by David Farber and Beth Bailey (New York: Columbia University Press, 2001), pp. 305–311

Mazur, Allan, 'The Journalists and Technology: Reporting about Love Canal and Three Mile Island', *Minerva*, 22.1 (1984), pp. 45–66

McCollum, Brett, 'Nuclear Energy is a Safe Industry', in *Nuclear Power: A Reference Handbook*, 2nd edn, ed. by Henderson, pp. 130–137

McKay, Alwyn, *The Making of the Atomic Age* (Oxford: Oxford University Press, 1984)

Medvedev, Zhores A., *Nuclear Disaster in the Urals*, trans. by George Saunders (London: Angus & Robertson, 1979)

Müller, Filip, *Eyewitness Auschwitz*, ed. by Susanne Flatauer (Chicago: Ivan R. Dee, 1999)

Mumford, Lewis, *The Myth of the Machine* (London: Secker and Warburg, 1971)

Murray, William, *Nuclear Turnaround: Recovery from Three Mile Island and the Lessons for the Future of Nuclear Power* ((n.p.): 1st Books Library, 2003)

Nelson, Joyce, *The Perfect Machine* (Ontario: Between the Lines, 1987)

Neufeld, Michael J., 'Introduction', in *Planet Dora: A Memoir of the Holocaust and the Birth of the Space Age*, by Yves Béon, trans. by Yves Béon and Richard L. Fague (Westview Press: Colorado, 1997), pp. ix–xxviii

'Nuclear Insurance and Disaster Relief', *nrc.org* (2014), available online at www.nrc.gov/reading-rm/doc-collections/fact-sheets/nuclear-insurance.pdf (accessed August 8, 2016)

Nuclear Power in Crisis, ed. by Andrew Blowers and David Pepper (London: Croom, 1987)

Nye, David, *American Technological Sublime* (Massachusetts: MIT, 1994)

O'Brian, John and Jeremy Borsos, *Atomic Postcards* (Chicago: Intellect, 2011)

O'Neill, William, *Coming Apart: An Informal History of America in the 1960s* (Times Books, 1971)

Oakley, Ronald J., *God's Country: America in the Fifties* (New York: Dembner Books, 1986)

Osif, Bonnie A., Anthony J. Baratta and Thomas W. Conkling, *TMI 25 Years Later: The Three Mile Island Power Plant Accident and Its Impact* (Pennsylvania: The Pennsylvania State University Press, 2004)

Our World or None, ed. by Dexter Masters and Katharine Way (Whittlesey House, 1946)

Pollock, Richard, 'Foreword', in *Three Mile Island: Turning Point*, by Bill Keisling (Washington: Express Publications, 1980), pp. 13–15

Prawitz, Jan, 'Is Nuclear Power Compatible with Peace? The Relation between Nuclear Energy and Nuclear Weapons', in *Facing up to Nuclear Power*, ed. by John Francis and Paul Abrecht (Philadelphia: The Westminster Press, 1976), pp. 102–109

Public Reactions to Nuclear Power, ed. by William R. Freudenburg and Eugene A. Rosa (Colorado: Westview Press, 1984)

Redniss, Lauren, *Radioactive: Marie & Pierre Curie. A Tale of Love and Fallout* (New York: HarperCollins, 2011)

Resistance to New Technology: Nuclear Power, Information Technology and Biotechnology, ed. by Martin Bauer (Cambridge: Cambridge University Press, 1997)

Rifas, Leonard, 'Cartooning and Nuclear Power: From Industry Advertising to Activist Uprising and Beyond', *PS: Political Science and Politics*, 40.2 (April, 2007), 255–260

Rosenthal, Peggy, 'The Nuclear Mushroom Cloud as Cultural Image', *American Literary History*, 3.1 (1991), pp. 63–92

Rotter, Andrew J., *Hiroshima: The World's Bomb* (Oxford: Oxford University Press, 2009)

Rudolf, Anthony, 'Primo Levi in London', in *The Voice of Memory*, by Primo Levi, ed. by Marco Belpoliti and Robert Gordon (Cambridge: Polity Press, 2001), pp. 23–33

Russ, Joanna, 'The Image of Women in Science Fiction', *Vertex* 32.1 (1974), pp. 53–57

Rydell, Robert W., *World of Fairs: The Century-of-Progress Expositions* (London: The University of Chicago Press, 1993)
Sanders, Jeff, 'Environmentalism', in *The Columbia Guide to America in the 1960s*, by David Farber and Beth Bailey (New York: Columbia University Press, 2001), pp. 273–281
Sanders, Ralph, 'Defense of Project Plowshare', *Technology and Culture*, 4.2 (1963), pp. 252–255
Schurr, Sam H., *et al.*, *Energy in America's Future* (London: John Hopkins University Press, 1979)
Schwebel, Milton, 'Effects of the Nuclear War Threat on Children and Teenagers: Implications for Professionals', *American Journal of Orthopsychiatry*, 52, pp. 608–618
Schwebel, Milton and Bernice Schwebel, 'Children's Reactions to the Threat of Nuclear Plant Accidents', *American Journal of Orthopsychiatry*, 51.26, 0–70
Scott, Alison M. and Christopher D. Geist, 'Preface', in *The Writing on the Cloud: American Culture Confronts the Atomic Bomb*, ed. by Alison M. Scott and Christopher D. Geist (Oxford: University Press of America, 1997), pp. v–1
Seitz, Ruth Hoover, *Susquehanna Heartland* (Harrisburg: RB books, 1992)
Sherwin, Martin J., *A World Destroyed: The Atomic Bomb and the Grand Alliance* (New York, 1977)
Smyth, H. D., *Atomic Energy For Military Purposes* (Pennsylvania: Maple Press 1945)
Sprod, Liam, *Nuclear Futurism* (Hants: Zero Books, 2012)
Steinbeck, John, *Travels with Charley: In Search of America* (London: Mandarin Paperbacks, 1991)
Steinmetz, Richard H. and Robert D. Hoffsommer, *This Was Harrisburg* (Pennsylvania: Stackpole Books, 1976)
Stocke, John Gregory, '"Suicide on the Instalment Plan:" Cold-War-Era Civil Defense and Consumerism in the United States', in *The Writing on the Cloud: American Culture Confronts the Atomic Bomb*, ed. by Alison M. Scott and Christopher D. Geist (Oxford: University Press of America, 1997), pp. 45–60
Stranahan, Susan Q., *Susquehanna, River of Dreams* (Maryland: Johns Hopkins University Press, 1993)
Susser, Mervyn, 'Three Mile Island Accident Continued: Further Comment, *Environmental Health Perspectives*, 105.6 (June 1997), pp. 566–567
Szasz, Ferenc Morton, *Atomic Comics* (Las Vegas: University of Nevada Press, 2012)
Talbott, Evelyn O., *et al.*, "Long-Term Follow-Up of the Residents of the Three Mile Island Accident Area: 1979–1988," *Environmental Health Perspectives* 111.3 (March 2003), pp. 341–348
—— 'Mortality among the Residents of the Three Mile Island Accident Area: 1979–1992', *Environmental Health Perspectives*, 108.6 (June 2000), pp. 545–552
Tanner, Thomas, '"The China Syndrome" as a Teaching Tool', *The Phi Delta Kappan*, 60.10 (June, 1979), pp. 708–712
Toffler, Alvin, *Future Shock* (New York: Bantam, 1971)
Valentine, John, *Atomic Crossroads: Before and After Sizewell* (London: Merlin Press, 1985)

Van Riper, A. Bowdoin, 'The Promise of Things to Come', in *Learning from Mickey, Donald and Walt. Essays on Disney's Entertainment Films*, ed. by A. Bowdoin Van Riper (North Carolina: McFarland, 2011), pp. 84–103

Vollmer, Richard, 'Representing the Nuclear Regulatory Commission', in *The Three Mile Island Nuclear Accident: Lessons and Implications*, ed. by Thomas H. Moss and David L. Sills (New York: The New York Academy of Sciences, 1981), pp. 110–113

Walker, J. Samuel, *Three Mile Island: A Nuclear Crisis in Historical Perspective* (London: University of California Press, 2004)

Wallis, John, 'Apocalypse at the Millennium', in *The End All Around Us: Apocalyptic Texts and Popular Culture*, ed. by John Wallis and Kenneth G. C. Newport (London: Equinox Publishing, 2009), pp. 71–97

Walsh, Edward J., *Democracy in the Shadows. Citizen Mobilization in the Wake of the Accident at Three Mile Island* (New York: Greenwood, 1988)

—— 'Resource Mobilization and Citizen Protest in Communities around Three Mile Island', *Social Problems*, 29.1 (October, 1981), pp. 1–21

Watts, Steven, *The Magic Kingdom: Walt Disney and the American Way of Life* (Missouri: University of Missouri Press, 1997)

Weaver, Roslyn, 'The Shadow of the End. The Appeal of Apocalypse in Literary Science Fiction', in *The End All Around Us: Apocalyptic Texts and Popular Culture*, ed. by John Wallis and Kenneth G. C. Newport (London: Equinox Publishing, 2009), pp. 173–198

Wing, Steve, 'Objectivity and Ethics in Environmental Health Science', *Environmental Health Perspectives*, 111.14 (November, 2003), pp. 1809–1818

Wing, Steve, *et al.*, 'A Reevaluation of Cancer Incidence near the Three Mile Island Nuclear Plant: The Collision of Evidence and Assumptions', *Environmental Health Perspectives*, 105.1 (January 1997), pp. 52–57

Wisnioski, Matthew, *Engineers for Change* (Cambridge: MIT, 2012)

Young, Bill and Bill Cotter, *Images of America: The 1964–1965 New York World's Fair* (Chicago: Arcadia, 2004)

Newspaper and Magazine

'1956: Queen Switches On Nuclear Power', *BBC News* (October 17, 1956), available online at http://news.bbc.co.uk/onthisday/hi/dates/stories/october/17/newsid_3147000/3147145.stm (accessed August 14, 2016)

'3-Mile Island Radiation Still Being Vented', *The Patriot*, March 29, 1979

'57 Incidents Like Harrisburg Reported in U.S.', *The Globe and Mail*, April 17, 1979

Ahearne, John F., 'Nuclear Power after Chernobyl', *Science*, New Series, 236.4802 (May 8, 1987), pp. 673–679

Allison, Wade, 'Viewpoint: We Should Stop Running Away from Radiation', *BBC* (March 26, 2011), available online at www.bbc.co.uk/news/world-12860842 (accessed May 15, 2016)

Alsop, Stewart, 'Eisenhower Pushes Operation Candor', *The Washington Post*, September 21, 1953

Anderson, Jack, 'Carter's Nuclear Team', *The Dispatch*, April 19, 1979
—— 'Three Mile Island Facts Revealed. Nuclear Probe Shocking', *Tuscaloosa News*, October 16, 1979
'Atom Power Plant Planned for Island in Pennsylvania', *New York Times*, February 12, 1967
Ayres, B. Drummond, 'Three Mile Island: Notes from a Nightmare; Three Mile Island: A Chronicle of the Nation's Worst Nuclear-Power Accident', *New York Times*, April 16, 1979
Benjamin, Stan, 'Three Mile Island and the Nuclear Power Industry', *Lewiston Journal*, March 17, 1980
Bernstein, Peter J., '"Fowl-Up" Sketched for Panel', *The Patriot*, March 30, 1979
Blackwelder, Brent, 'American Forum: Nuclear Reactor Roulette', *The Phoenix News* (April 30, 2011), available online at www.phoenixvillenews.com/article/20110430/TMP06/304309971 (accessed May 12, 2015)
Bower, Bradley C., 'TMI anniversary', *The Times* (March 28, 2011), available online at www.timesonline.com/tmi-anniversary/image_ec8fc588-5963-11e0-b6d7-0017a4a78c22.html (accessed July 13, 2016)
Bradley, Mary O., Don Sarvey and Terry Williamson, 'Leak Poses "No Danger" to Populace', *The Evening News*, March 28, 1979
Branson, Constance Y., 'Legacy Seen in TMI: Public Demand to Be Fully Informed', *The Patriot*, May 16, 1979
Brutto, Carmen, 'Nuclear Dangers Doubted', *The Patriot*, March 30, 1979
Buchwald, Art, 'Way Wind Is Blowing Shapes Nuclear Views', *The Scranton Times*, April 13, 1979
Bukro, Casey, 'Three Mile Island: Answers Are All 'Maybes'', *The Gazette*, October 31, 1979
Burnham, David, 'But Does it Satisfy Nuclear Experts?', *New York Times*, March 18, 1979
—— 'C.I.A. Papers, Released to Nader, Tell of 2 Soviet Nuclear Accidents', *New York Times*, November 26, 1977
—— 'U.S. Proposes to Fine Utility for Not Keeping Unstable Ex-Employee Out of Nuclear Plant', *New York Times*, March 30, 1976
Butt, Ned M., 'Nuclear Plants Can Be Made Safe', *The Patriot*, May 10, 1979
Caldicott, Helen, 'Unsafe at Any Dose', *New York Times* (April 30, 2011), available online at www.nytimes.com/2011/05/01/opinion/01caldicott.html?_r=0 (accessed July 21, 2016)
Canby, Vincent, 'China Syndrome is First-Rate Melodrama . . .', *New York Times*, March 18, 1979
Cannella, Anthony, 'Crash Data Sought by Nuclear Regulators', *The Scranton Times*, March 30, 1979
—— 'Spurred by Nuclear Mishap, Foes Want Plant Stopped', *The Sunday Times, Scranton*, April 1, 1979
Cannon, Lou, 'Actor, Governor, President, Icon', *The Washington Post*, June 6, 2004
'Carter Acts on TMI: A Flood of Changes', *Science News*, 116.24 (December 15, 1979)

Chamberlain, John, 'Three Mile Island Perspective Lost', *Daily News*, Kentucky, May 2, 1979

'A Cloud of Gloom and Doom', *New Scientist*, April 23, 1987

'Colossal H-Bomb Hole', *LIFE*, June 21, 1963

Crable, Ad, 'TMI 25 Years After: Security Now Top Worry', *LancasterOnline* (March 19, 2004), available online at http://lancasteronline.com/news/tmi-years-after-security-now-top-worry/article_563825a4-a986-561c-ba87-d0f38ab88844.html (accessed March 18, 2016)

Cress, Joseph, 'Area Residents Remember Midstate During TMI Partial Meltdown', *The Sentinel* (March 23, 2014), available online at http://cumberlink.com/news/local/history/area-residents-remember-midstate-during-tmi-partial-meltdown/article_c0dd3214-b2e2-11e3-bed9-0019bb2963f4.html (accessed July 14, 2016)

Crow, Margie, 'Nukes vs. Anti-Nukes: Malignant Monster Meets Critical Mass Movement', *Off Our Backs*, 9.5 (May 1979)

'D'oh! German Broadcaster Pulls "Nuclear" Simpsons Episodes', *New York Post* (March 28, 2011), available online at http://nypost.com/2011/03/28/doh-german-broadcaster-pulls-nuclear-simpsons-episodes/ (accessed May 16, 2016)

DeAndrea, Francis T., 'Evacuees Wonder if Life Will Ever Be the Same', *The Sunday Times, Scranton*, April 1, 1979

'Demonstrators in Japan Urge Nuclear Shutdowns', *The Scranton Times*, April 2, 1979

'Deposition In Plutonium Case is Read', *The Patriot*, March 29, 1979

Dickson, David, 'Britain Chooses U.S.-Designed Reactor', *Science*, New Series, 235.4789 (February 6, 1987)

Dorman, William A. and Daniel Hirsch, 'The U.S. Media's Slant', in *Chernobyl: The Emerging Story, The Bulletin of the Atomic Scientists*, ed. by Harrison Brown (August/September 1986), 54–56

Duscha, Julius, 'The Media and Three Mile Island', *The Baltimore Sun*, April 19, 1979

Dvorchak, Bob and Harry F. Rosenthal, 'Seven Days at Three Mile Island Nuclear Power Plant', *Daily News*, April 8, 1979

Fackler, Martin, 'Japan's Nuclear Energy Industry Nears Shutdown, at Least for Now', *New York Times*, (March 8, 2012), available online at www.nytimes.com/2012/03/09/world/asia/japan-shutting-down-its-nuclear-power-industry.html?_r=2&emc=tnt&tntemail1=y (accessed July 17, 2016)

—— 'Radiation Fears and Distrust Push Thousands from Homes', *The New York Times* (March 17, 2011), available online at www.nytimes.com/2011/03/18/world/asia/18displaced.html?pagewante%20d=all%3E&_r=0 (accessed July 18, 2016)

'Few Remaining Parishioners Think About "Living with Risk"', *The Scranton Times*, April 2, 1979

Fishlock, Diana, 'TMI Stories: Dina Gonzalez, Scary Time for a 6-year-old', *The Patriot* (March 22, 2009), available online at www.pennlive.com/specialprojects/index.ssf/2009/03/tmi_stories_dina_gonzalez.html (accessed November 11, 2015)

—— 'TMI Stories: Ken Miller, Wrote Medical Emergency Plan', *The Patriot* (March 22, 2009), available online at www.pennlive.com/specialprojects/index.ssf/2009/03/tmi_stories_ken_miller.html (accessed November 11, 2015)

Flannery, Joseph X., 'Some Experts Noncommittal', *The Scranton Times*, March 30, 1979

Frantz, Jeff, 'Experts Debate Three Mile Island's Safety Buffer', *Pennlive*, (March 19, 2011), available online at www.pennlive.com/midstate/index.ssf/2011/03/experts_debate_three_mile_isla.html (accessed November 11, 2015)

'General Public Utilities Units Receive Approval to Build Nuclear Plant', *Wall Street Journal*, November 3, 1969

Gilbert Associates, 'Get Involved', *Delaware County Daily Times* (June 12, 1967)

Gilliland, Donald, 'For Some, Japanese Nuclear Emergency Mirrors 1979's Three Mile Island', *Pennlive* (March 12, 2011), available online at www.pennlive.com/midstate/index.ssf/2011/03/for_some_japanese_nuclear_emer.html (accessed May 17, 2015)

Graff, Don, 'Three Mile Island Revisited', *Daily News*, November 7, 1979

Greenpeace, 'An American Chernobyl: Nuclear "Near Misses" at U.S. Reactors Since 1986', *Greenpeace.org* (2006), available online at www.greenpeace.org/wp-content/uploads/legacy/Global/usa/report/2007/9/an-american-chernobyl-nuclear.pdf (accessed July 30, 2016)

'Harts Give Insights on Homesteading at Festival', *The Gazette*, Wellsboro, July 25, 1979

Harwood, Jon, 'Neighbors Vow To Bury N-Plant', *The Patriot*, May 22, 1979

Hines, William, 'Still Plenty of Fallout Over '50s A-Bomb Tests. When Science Prostituted, Everyone Loses', *The Patriot*, May 1, 1979

'Hiroshima-Nagasaki Week', *Peace Newsletter*, August 1980

Howland, Barker, '"The China Syndrome" Attendance is Up', *The Patriot*, March 30, 1979

—— 'Three Mile Island: Legacy of Trouble', *The Patriot*, March 29, 1979

Hume, Ellen, 'Three Mile Island Study: "Amazed" Housewife on Blue-Ribbon Panel', *Los Angeles Times*, April 12, 1979

Hunt, Marilyn Scranten, 'Nuclear Power a Pandora's box', The Forum, *Pittsburgh Catholic*, May 1, 1981

'If The Public Knew. . .Nuclear Power Plants?', *Delaware County Daily Times*, September 4, 1973

'Inhabitants Wonder What to Believe', *WashingtonPost.com* (1979), available online at www.washingtonpost.com/wp-srv/national/longterm/tmi/stories/ch14.htm (accessed July 27, 2016)

The Inquirer Editorial Board, 'How not to build trust in nuclear power', *The Inquirer Daily News* (November 27, 2009), available online at www.philly.com/philly/blogs/inq_ed_board/How_not_to_build_trust_in_nuclear_power.html?c=r (accessed July 22, 2016)

'Island Scene of Blaze', *The Record-Angus*, January 20, 1971

Jansen, Donald, 'Radiation Is Released in Accident at Nuclear Plant in Pennsylvania', *New York Times*, March 29, 1979

Johnson, James H. and Donald J. Zeigler, 'Distinguishing Human Responses to Radiological Emergencies', *Economic Geography*, 59.4 (October, 1983), 386–402

Kilpatrick, James J., 'Three Mile Island—A Calm perspective', *The Deseret News*, April 4, 1980

Kiner, Deb, 'Elks Theatre Will Mark Tmi Anniversary with "My Father, The Old Horse" and "The China Syndrome"', *Pennlive* (May 5, 2014), available online at www.pennlive.com/entertainment/index.ssf/2014/05/elks_theatre_will_mark_tmi_ann.html (accessed August 25, 2016)

King, Wayne, 'Concern Rises in South Carolina, Home of Many Nuclear Reactors; In Spite of Assurances and Plans for Safety Check, Worry Grows', *New York Times*, April 1, 1979

Kirkpatrick, Rich, 'Concerns Linger Ten Years After Three Mile Island', *The Dispatch*, March 25, 1989

Klaus, Mary, 'Decision on Reporting is Political, Speaker Says. Air TMI Views, Public Urged', *The Patriot*, May 1, 1979

—— 'Radiation Above Normal: Scientists Seek Closing', *The Patriot*, March 30, 1979

Knight, Chris, 'Three Mile Island Tour', *The Patriot* (September 8, 2010), available online at http://videos.pennlive.com/patriot-news/2009/10/three_mile_island_tour.html (accessed August 5, 2015)

LaBelle, G. G., 'Nuclear Energy Concern Sweeps Nation', *The Sunday Times*, Scranton, April 1, 1979

Lacey, Hal, 'Nuclear Mishap Called "Minor" by Officials', *The Scranton Times*, March 30, 1979

Langer, Elinor, 'Project Plowshare: AEC Program for Peaceful Nuclear Explosives Slowed Down by Test Ban Treaty', *Science, New Series*, 143.3611 (1964), pp. 1153–1155

'Lawmakers Privately Briefed at TMI', *The Patriot*, Harrisburg, March 30, 1979

LeDoux, Jerome, 'All in Perspective', *Pittsburgh Catholic*, July 6, 1979

Lenton, Garry, 'Poll Shows Midstate Supports Nuclear Power', *PennLive* (September 8, 2008), available online at www.pennlive.com/midstate/index.ssf/2008/09/midstate_supports_nuclear_powe.html (accessed March 17, 2016)

Life, May 1979, 2.5

Lohr, Steve, 'Britain Suppressed Details of '57 Atomic Disaster', *The New York Times* (January 2, 1988), available online at www.nytimes.com/1988/01/02/world/britain-suppressed-details-of-57-atomic-disaster.html (accessed August 30, 2016)

MacLeod, Scott, 'Life Not Back to Normal after Three Mile Island Mishap', *The Bryan Times*, October 1, 1979

Macleod, Gordon K., 'Some Public Health Lessons from Three Mile Island: A Case Study in Chaos', *Ambio*, 10.1 (1981), 18–23

Mansfield, Duncan, 'Nuclear Revival Begins with the Restart of TVA Reactor', *The Times* (May 5, 2007), available online at www.timesonline.com/nuclear-revival-begins-with-restart-of-tva-reactor/article_1695aa11-ae89-5dc3-ba0c-292c37a3dfa4.html (accessed July 27, 2016)

Marshall, Eliot, 'Investigation Confirms TMI Cleanup Problems', *Science,* New Series, 221.4618 (September 30, 1983)
——'NRC Must Weigh Psychic Costs', *Science,* New Series, 216.4551 (June 11, 1982), pp. 1203–1204
mc, 'the china syndrome', *Off Our Backs*, 9.5, May 1979
'Mental Problems Are Expected for Residents in N-Plant Area', *The Scranton Times*, April 3, 1979
Meyers, John A., 'A Letter from the Publisher', *Time*, April 9, 1979, 113.15
Miller, Dan, 'NRC Tuesday to Hold Open House Regarding Three Mile Island in Londonderry Township', *The Patriot* (April 7, 2010), available online at www.pennlive.com/midstate/index.ssf/2010/04/nrc_tuesday_to_hold_open_house.html (accessed June 19, 2016)
Miller, Peter, 'Susquehanna: America's Small-Town River', *National Geographic*, 167.3, (March 1985), pp. 352–383
'N-Plant Builder is Sued', *The Patriot*, May 1, 1979
Navias, Geoffrey, 'Three Movements: Strategies for the 1980's', *Peace Newsletter,* August 1980
'Not Prepared for Mishap, Operator of N-Plant Admits', *The Scranton Times*, April 13, 1979
'Nuclear Energy Fight Finally Goes Public', *The Scranton Times*, April 3, 1979
'Nuclear Morality', *Pittsburgh Gazette*, April 6, 1979
'A Nuclear Nightmare', *Time*, April 9, 1979, 113.15
'Nuclear Plant Breakdown Releases Radiated Steam', *The Scranton Times*, March 28, 1979
'Nuclear Protest: The Darker Side of Technology', *The Patriot*, May 8, 1979
'Penelec Forecasts Steady Growth', *Wellsboro Gazette*, January 25, 1978
'Penelec Plant Ranked No.1', *Wellsboro Gazette*, September 22, 1976
Perlman, Fredy, 'Progress and Nuclear Power: The Destruction of the Continent and Its Peoples', *Fifth Estate*, (April 8, 1979), available online at https://archive.org/details/ProgressAndNuclearPowerTheDestructionOfTheContinentAndIts Peoples (accessed May 13, 2015)
Peterson, Alfred L., 'Three Mile Island and the Military, *The Baltimore Sun*, April 9, 1979
Pietila, Antero, 'Distrust, Frustration Pervade Area Near A-plant as Some Prepare to Leave', *The Baltimore Sun*, March 31, 1979
Projansky, Carolyn, J., *et al.*, 'Nuclear Power is a Feminist Issue', *Off Our Backs*, 9.5 (May 1979)
'Pump Breaks Down. Radioactive Steam Escapes in the Air', *The Tuscaloosa News*, March 28, 1979
Quigley, Roger, 'Goldsboro: Tranquility and Anger', *The Patriot*, March 29, 1979
'Radiation Dose Revised Upwards', *The Patriot*, May 4, 1979
'Radiation Leaks from Nuclear Plant Continue', *The Scranton Times*, March 29, 1979
'Radiation: Who Said What', *The Patriot*, Harrisburg, March 30, 1979
Richman, Alan, '"Good News" at Three Mile Island, But 1,000 Stage Harrisburg Protest', *The New York Times*, April 9, 1979

Rieders, Clifford A., 'Conservation: Maybe the Tree Huggers Were Right', *Phoenix News* (March 26, 2011), available online at www.phoenixvillenews.com/article/20110326/TMP06/303269970 (accessed April 17, 2016)

Ritzel, Rebecca J., 'TMI a Look Back', *LancasterOnline* (March 24, 2004), available online at http://lancasteronline.com/news/tmi-a-look-back/article_2e446a14-d95e-53a4-959b-02eb042ab70f.html (accessed April 20, 2016)

Roberts, Richard, '4 Counties Still on Alert', *The Patriot*, March 30, 1979

—— 'Panel Probes Health Aspects of TMI Accident', *The Patriot*, May 23, 1979

—— 'TMI Startup "Blacklash" Seen', *The Patriot*, May 11, 1979

Robinson, Diana, 'Where Were You When Three Mile Island Had Its Partial Meltdown? Readers Share Their Stories', *PennLive* (March 26, 2014), available online at www.pennlive.com/midstate/index.ssf/2014/03/where_were_you_when_three_mile.html (accessed April 12, 2015)

Saxon, Wolfgang, 'G. C. Minor, 62, an Engineer Who Criticized Nuclear Power', *The New York Times* (July 31, 1999), available online at www.nytimes.com/1999/07/31/us/g-c-minor-62-an-engineer-who-criticized-nuclear-power.html (accessed August 1, 2016)

'Sellafield Comment "Offensive"', *BBC* (May 30, 2003), available online at http://news.bbc.co.uk/1/hi/northern_ireland/2949334.stm (accessed August 30, 2016)

Shalett, Sidney, 'First Atomic Bomb Dropped on Japan; Missile Is Equal to 20,000 Tons of TNT; Truman Warns Foe of a "Rain of Ruin"', *The New York Times*, August 7, 1945

'Silkwood Case Laid to Rest', *Science News*, 130.9 (August 30, 1986)

Smart, Gil, 'The Media and the Meltdown', *LancasterOnline* (March 20, 2011), available online at http://lancasteronline.com/opinion/the-media-and-the-meltdown/article_d78417a2-b901-5a16-b2f9-b714dfc6d64f.html (accessed July 19, 2016)

Stuhler, Leonard, 'To the Editor', *Wellsboro Gazette*, April 18, 1979

'Susquehanna Island To Be A-Plant Site', The *Washington Post, Times Herald*, February 12, 1967)

Thompson, Helen, 'Chernobyl's Bugs: The Art and Science of Life After Nuclear Fallout', *Smithsonian Magazine* (April 26, 2014), available online at www.smithsonianmag.com/arts-culture/chernobyls-bugs-art-and-science-life-after-nuclear-fallout-180951231/?no-ist (accessed May 18, 2016)

'Three Mile Island Intensifies Nuclear Power Controversy', *The Hartford Courant*, September 28, 1979

'Three Mile Island Meltdown Was Near', *Observer-Reporter*, January 25, 1980

'Timeline: Japan power plant crisis', *BBC* (March 13, 2011), available online at www.bbc.co.uk/news/science-environment-12722719 (accessed April 19, 2016)

'TMI Anniversary Through the Eyes of Midstate Resident', *The Sentinel* (March 27, 2014), available online at http://cumberlink.com/news/local/capital_region/tmi-anniversary-through-the-eyes-of-midstate-resident/article_23dd07dc-b616-11e3-a0a9-0019bb2963f4.html (accessed July 19, 2016)

Toland, Bill and Clif Page, 'Ghosts of Chernobyl', *The Times*, December 15, 2002

Tomb, Gordon, 'BP Disaster Echoes TMI Incident—What Have We Learned?', *PennLive* (June 20, 2010), available online at www.pennlive.com/editorials/index.ssf/2010/06/bp_diaster_echoes_tmi_incident.html (accessed June 12, 2016)

Topham, Laurence, Alok Jha and Will Franklin, 'Building the Bomb', *The Guardian* (September 22, 2015), available online at www.theguardian.com/us-news/ng-interactive/2015/sep/21/building-the-atom-bomb-the-full-story-of-the-nevada-test-site (accessed August 8, 2016)

'Trouble at Ground Zero', *The Gettysburg Times*, April 2, 1979

'Uncommon Sense', *Wellsboro Gazette*, April 11, 1979

'Voice of the People: A Sickening Way to Belittle Fears over Sellafield', *The People*, June 1, 2003

Von Dobeneck, Monica, 'TMI stories: John Milkovich', *Pennlive* (March 22, 2009), available online at www.pennlive.com/specialprojects/index.ssf/2009/03/tmi_stories_john_milkovich.html (accessed July 19, 2016)

Wasserman, Harvey, 'People Were Killed by Three Mile Island and Other Nuclear Disasters', *OpEdNews* (November 19, 2016), available online at www.opednews.com/articles/genera_harvey_w_071119_people_were_killed_b.htm (accessed April 13, 2016)

'We All Live in Harrisburg', *The Economist*, April 7, 1979

Wenner, David, 'TMI Will Be First U.S. Power Plant Formally Tested on Terrorist Attack Plan', *PennLive* (April 16, 2013), available online at www.pennlive.com/midstate/index.ssf/2013/04/tmi_three_mile_island_drill_mi.html (accessed July 15, 2016)

'What is Radiation?' *The Scranton Times*, March 28, 1979

Willard, Hal, 'Atomic Plants Increase in Chesapeake Bay Area', *The Washington Post, Times Herald*, November 6, 1969

'Women, Kids Evacuated', *The Scranton Times*, March 30, 1979

Yoffe, Emily, 'Pennsylvanians Lobby for N-Shutdown', *The Patriot*, May 8, 1979

Official Documents, Hearings, Speeches, and Papers

'The Atomic Energy Act', United States Nuclear Regulatory Commission, August 30, 1954

Atoms for Peace? (Commonwork Land Trust Information Service on Energy: 1982)

Eisenhower, Dwight D., 'Atoms for Peace', 470th Plenary Meeting of the United Nations General Assembly, 8 December 1953, in *IAEA*, available online at https://www.iaea.org/about/history/atoms-for-peace-speech (accessed July 14, 2016)

Gosling, F. G. and The United States Department of Energy, *The Manhattan Project: Making the Atomic Bomb* (Washington: Department of Energy, 1999)

Houts, Peter S., *et al.*, *Health-Related Behavioral Impact of the Three Mile Island Nuclear Incident*, 'TMI Advisory Panel on Health Research Studies of The Pennsylvania Department of Health', The Pennsylvania State University, College of Medicine, and The Pennsylvania Department of Health (Pennsylvania: April 8, 1980), Part 1

'IAEA General Conference Adopts Resolution on the Physical Protection of Nuclear Material and Nuclear Facilities', *IAEA: International Atomic Energy Agency* (September 21, 2001), available online at https://www.iaea.org/newscenter/press releases/iaea-general-conference-adopts-resolution-physical-protection-nuclear-material-and-nuclear-facilities (accessed July 20, 2016)

Impact Abroad of the Accident at the Three Mile Island Nuclear Power Plant, 'Subcommittee on Energy, Nuclear Proliferation and Federal Services of the Committee on Governmental Affairs', United States Senate (Washington: Library of Congress, May 1980)

INPO, 'Home', *INPO* (2015), available online at www.inpo.info/Index.html (accessed July 18, 2016)

Kemeny, John G., *et al.*, *The Need for Change: The Legacy of TMI*, 'Report of the President's Commission on the Accident at Three Mile Island' (Washington D.C., October 1979)

Kennedy, John F., 'Acceptance Speech', Hyannis Armory, Hyannis, Massachusetts, November 9, 1960, in *John F. Kennedy: Presidential Library and Museum* (n.d.), available online at https://www.jfklibrary.org/Research/Research-Aids/JFK-Speeches/Hyannis-MA-Acceptance-Speech_19601109.aspx (accessed August 12, 2012)

Kennedy, John F., 'Address Before the 18th General Assembly of the United Nations', United Nations, 20 September 1963, in *John F. Kennedy: Presidential Library and Museum* (n.d.), available online at www.jfklibrary.org/Research/Ready-Reference/JFK-Speeches/Address-Before-the-18th-General-Assembly-of-the-United-Nations-September-20–1963.aspx (accessed August 12, 2012)

Kennedy, John F., 'Address of Senator John F. Kennedy Accepting the Democratic Party Nomination for the Presidency of the United States', Memorial Coliseum, Los Angeles, July 15, 1960, in Gerhard Peters and John T. Woolley, *The American Presidency Project* (2016), available online at www.presidency.ucsb.edu/ws/?pid=25966 (accessed July 16, 2016)

Kennedy, John F., 'Remarks at the Dedication of the Aerospace Medical Health Center', San Antonio, Texas, November 21, 1963, in *John F. Kennedy: Presidential Library and Museum* (n.d.), available online at https://www.jfklibrary.org/Research/Research-Aids/JFK-Speeches/San-Antonio-TX_19631121.aspx (accessed August 12, 2012)

Kurokawa, Kiyoshi, *et al.*, *The National Diet of Japan. The Official Report of The Fukushima Nuclear Accident Independent Investigation Commission* (2012), available online at https://www.nirs.org/fukushima/naiic_report.pdf (accessed August 14, 2016)

McClure, James A., *et al.*, *The Chernobyl Accident*, 'Hearing before the Committee on Energy and Natural Resources', United States Senate, Ninety-Ninth Congress, Second Session on the Chernobyl Accident and Implementations for the Domestic Nuclear Industry (Washington D.C., June 19, 1986)

Nuclear Powerplant Safety After Three Mile Island, 'Report Prepared by the Subcommittee on Energy Research and Production of the Committee on Science and Technology', United States House of Representatives, Ninety-Sixth Congress, Second Session (Washington D.C., March 1980)

Nuclear Safety—Three Years After Three Mile Island, 'Joint Hearing before the Subcommittees of the Committees on Government Operations and Interior and Insular Affairs', House of Representatives, Ninety-Seventh Congress, Second Session (Washington: March 12, 1982)

Order Suspending Power Operation and Order to Show Cause (Effective Immediately), United States Nuclear Regulatory Commission (Washington D.C., March 12, 1987)

'Pressurized Water Reactors', *United States Nuclear Regulatory Commission* (2015), available online at www.nrc.gov/reactors/pwrs.html (accessed July 15, 2016)

'Project "Candor"', Security Information Secret, Unclassified (July 22, 1953), in *The Dwight D. Eisenhower Library* (n.d.), available online at https://www.eisenhower.archives.gov/research/online_documents/atoms_for_peace/Binder17.pdf (accessed July 19, 2016)

Rogovin, Mitchell, *et al.*, *Volume 1. Three Mile Island. A Report to the Commissioners and to the Public*, Nuclear Regulatory Commission Special Inquiry Group (Washington D.C., 1980)

Smyth, H. D., *A General Account of the Development of Methods of Using Atomic Energy for Military Purposes under the Auspices of the United States Government 1940–1945* (Washington D.C., 1945)

Staff Studies, Nuclear Accident and Recovery at Three Mile Island, 'Subcommittee on Nuclear Regulation for the Committee on Environment and Public Works', United States Senate, Ninety-Sixth Congress, Second Session (Washington: July 1980)

Three Mile Island Nuclear Powerplant Accident, 'Hearings before the Subcommittee on Nuclear Regulation of the Committee on Environment and Public Works', United States Senate, Ninety-Sixth Congress, First Session (Washington: April 10, 28, and 30, 1979), Part 1

Treaty Banning Nuclear Weapon Tests in the Atmosphere, in Outer Space and Under Water, Moscow, October 15, 1963

Truman, Harry S., 'Statement by the President of the United States', Washington D.C., The White House, August 6, 1945, in *Harry S. Truman Library & Museum* (2016). https://www.trumanlibrary.org/whistlestop/study_collections/bomb/large/documents/index.php?documentid=59&pagenumber=1 (accessed July 19, 2016)

Udall, Morris K., *et al.*, *Accident at the Three Mile Island Nuclear Powerplant*, 'Oversight Hearing before the Subcommittee on Energy and the Environment of the Committee on Interior and Insular Affairs, House of Representatives', Ninety-Sixth Congress, First Session (Washington D.C., May 21, 24, 1979), Part II

United States Nuclear Regulatory Commission, 'Backgrounder: Three Mile Island Accident' (2013), available online at www.nrc.gov/reading-rm/doc-collections/fact-sheets/3mile-isle.pdf (accessed August 2, 2016)

'United States Nuclear Tests July 1945 through September 1992', United States Department of Energy Nevada Operations Office (Nevada, 2000)

United States Strategic Bombing Survey: The Effects of the Atomic Bombings of Hiroshima and Nagasaki (June 19, 1946), 1–51, in *Harry S. Truman Library & Museum* (2016), available online at www.trumanlibrary.org/whistlestop/study_collections/bomb/large/documents/index.php?documentdate=1946–06–19&documentid=65&studycollectionid=abomb&pagenumber=1 (accessed July 5, 2016)

'USAEC General Advisory Committee Report on the "Super", October 30, 1949', in *The American Atom. A Documentary History of Nuclear Policies from the Discovery of Fission to the Present. 1939–1984*, ed. by Robert Williams and Philip Cantelon (Philadelphia: University of Pennsylvania Press, 1984), pp. 120–128

Lectures, Dissertations, and Conferences

Bisconti, Ann, 'United States Public Opinion About Nuclear Energy: Past, Present and Future', *From Its Birthplace: A Symposium on the Future of Nuclear Power* (University of Pittsburgh, March 27–28, 2012), available online at www.thornburghforum.pitt.edu/sites/default/files/Nuclear%20Symposium%20report%20FINAL%20report%2011_5_12.pdf (accessed May 22, 2016) pp. 37–38

Blagg-Miller, Penny, 'Unconscious Fantasies of Children Living in the Three Mile Island Area' (doctoral dissertation, The Union Graduate School, 1982)

Caldicott, Helen, 'Dr. Helen Caldicott: What We Learned from Fukushima', Seattle Community Media (2012), available online at http://seattlecommunitymedia.org/node/30758 (accessed July 17, 2016)

Denton, Harold, 'Nuclear Power Accidents', *Knox County Public Library* (April 29, 2011), available online at www.knoxlib.org/about/news-and-publications/podcasts/brown-bag-green-book-podcast/nuclear-power-plant-accidents (accessed July 30, 2016)

Fuller, Loïe, 'Lecture on Radium' (London, January 20, 1911)

Moss, H., 'Background of the Three Mile Island Nuclear Accident, I: General Discussion', in *The Three Mile Island Nuclear Accident: Lessons and Implications*, ed. by Thomas H. Moss and David L. Sills (New York: The New York Academy of Sciences, 1981), pp. 48–53

Ogley-Oliver, Emma, 'Development of Activism: The Elders of the Anti-Nuclear Movement' (doctoral dissertation, Georgia State University, 2012)

Upton, Arthur C., 'Heath Impact of the Three Mile Island Accident', in *The Three Mile Island Nuclear Accident: Lessons and Implications*, ed. by Thomas H. Moss and David L. Sills (New York: The New York Academy of Sciences, 1981), pp. 63–75

Vollmer, Richard, 'Representing the Nuclear Regulatory Commission', in *The Three Mile Island Nuclear Accident: Lessons and Implications*, ed. by Thomas H. Moss and David L. Sills (New York: The New York Academy of Sciences, 1981), pp. 110–113

Von Burg, Ron, 'The Cinematic Turn in Public Discussions of Science' (doctoral dissertation, University of Pittsburgh, 2005)

Letters

Einstein, Albert, 'Letter to F. D. Roosevelt', August 2, 1939, in *AtomicArchive*. www.atomicarchive.com/Docs/Begin/Einstein.shtml (accessed August 18, 2016)
Kennedy, J. F., 'A Message to You from the President', *Life*, September 15, 1961
Szilard, Leo, 'Leo Szilard and the Discovery of Fisson', in *The American Atom. A Documentary History of Nuclear Policies from the Discovery of Fission to the Present. 1939–1984*, ed. by Robert Williams and Philip Cantelon (Philadelphia: University of Pennsylvania Press, 1984), pp. 7–12

Interviews

Clark, Adam Scott, on *Beaver Pig* (email to Grace Halden, February 15, 2016, 21:12 p.m.)
Noble, Phil, interview with Grace Halden, EDF Energy Nuclear, June 2, 2015
Tucker, Colin, interview with Grace Halden, Sizewell B, April 13, 2015

Film and Television

A is for Atom, dir. by Carl Urbano (General Electric Company, 1952)
'The Atomic Artists', dir. by Emily Taguchi (PBS, 2011), in *Frontline* (2014), available online at www.pbs.org/wgbh/pages/frontline/the-atomic-artists/?autoplay (accessed May 25, 2016)
Atomic City, dir. by Jerry Hopper (Paramount, 1952)
Atomic Energy as a Force for Good, dir. by Robert Stevenson (The Christophers, 1955)
Atomic Power at Shippingport (Westinghouse Electric Corp, c. 1950)
Atomic Twister, dir. by Bill Corcoran (TBS, 2002)
Balionis, Amanda and Paul Franz, 'TALKING POINTS: More Nukes?', *Lancaster Online* (April 12, 2009), available online at http://lancasteronline.com/news/talking-points-more-nukes/article_8f797de0-0f4f-58b0-85b2-aaf9fa47d01c.html (accessed April 19, 2016)
The Beginning or the End, dir. by Norman Taurog (Metro-Goldwyn-Mayer, 1947)
Chernobyl Heart, dir. by Maryann DeLeo (HBO, 2003)
The China Syndrome, dir. by James Bridges (Columbia Pictures, 1979)
The China Syndrome Trailer, dir. by James Bridges (Columbia Pictures, 1979)
'Crepes of Wrath', *The Simpsons*, Fox, April 15, 1990
Dark Circle, dir. by Judy Irving (Independent Documentary Group, 1982)
The Day After, dir. by Nicholas Meyer (ABC, 1983)
Deadly Deception: General Electric, Nuclear Weapons and Our Environment, dir. by Debra Chasnoff (GroundSpark, 1991)
Def-Con 4, dir. by Paul Donovan (New World Pictures, 1985)
Dr Strangelove or: How I Learned to Stop Worrying and Love the Bomb, dir by. Stanley Kubrick (Columbia Pictures, 1964)
Duck and Cover, Official Civil Defense Film (Archer Productions, 1951)
Face the Nation, CBS, April 1, 1979
Fallout, Office of Civil and Defense Mobilization (Creative Arts Studio, 1955)

Five, dir. by Arch Oboler (Columbia Pictures, 1951)
Heavy Water: A Film for Chernobyl, dir. by David Bickerstaff and Phil Grabsky (Seventh Art, 2006)
'Homer Defined', *The Simpsons*, Fox, October 17, 1991
'Homerland', *The Simpsons*, Fox, September 29, 2013
Inviting Disaster: Three Mile Island, dir. by David DeVries (History Channel, 2003)
'It's Electric!', *The Atom for Peace*, c. 1950(58?)
Jarriel, Tom, 'Three Mile Island', *ABC News*, ABC, March 30, 1979
'Karen Silkwood: A Life on the Line', host, Harry Smith (*Towers Productions*, 2001)
'King Size Homer', *The Simpsons*, Fox, November 5, 1995
'Last Exit to Springfield', *The Simpsons*, Fox, 11 March 1993
The Medical Aspects of Nuclear Radiation, U.S. Air Force (Cascade Pictures of California, c. 1950)
A Navajo Journey, dir. by C. J. Colby (Kerr-McGee, 1952)
Our Friend the Atom, dir. by Hamilton Luske (Walt Disney Productions, 1957)
Pandora's Promise, dir. by Robert Stone (Impact Partners, 2013)
Panic in Year Zero! dir. by Ray Milland (American International Pictures, 1962)
Plowshare, United States Atomic Energy Commission (San Francisco: W. A. Palmer Films, (n.d.))
Power and Promise: The Story of Shippingport, The United States Atomic Energy Commission (Mode-Art Pictures, c. 1950)
Powering America, dir. by Stephen Vidano (The Heritage Foundation, 2012)
Rocketship XM, dir. by Kurt Neumann (Lippert Pictures, 1950)
'Shelter Skelter', *The Twilight Zone*, CBS, 21 May 1987
Silkwood, dir. by Mike Nichols
Special Bulletin, dir. by Edward Zwick (NBC, 1983)
Super 8, dir. by J. J. Abrams (Paramount Pictures, 2011)
Survival Under Atomic Attack, Official United States Civil Defense Film (Castle Films, 1951).
Surviving the Tsunami: My Atomic Aunt, dir. by Kyoko Miyake (NHK, 2013)
Testament, dir. by Lynne Littman (Paramount, 1983)
Them!, dir. by Gordon Douglas (Warner Bros, 1954)
'Two Cars in Every Garage and Three Eyes in Every Fish', *The Simpsons*, Fox, November 1, 1990
Unknown World, dir. by Terry O. Morse (Lippert Pictures, 1951)
Victory Through Airpower, dir. by Perce Pearce (Walt Disney Productions, 1943).
When the Wind Blows, dir. by Jimmy Murakami (Kings Road Entertainment, 1986)
White Horse, dir. by Maryann DeLeo (HBO, 2008)

Literature and Comics and Poetry

All-Atomic Comics, 1 (EduComics, 1979).
'Bill Cosmo and the Plutonium Pile', *Future World Comics*, 1 (New York: Summer 1946)

'The Bomb that Won the War', *Science Comics*, 1, (Springfield: Humor Publications, January 1946)
Brinkley, William, *The Last Ship* (New York: Viking Press, 1988)
Clark, Adam Scott, *Beaver Pig* (Lancaster: unpublished manuscript, 2013)
'Could Science Blow the World Apart?', *Picture News*, 1.1. (New York: January 1946)
'How Atomic Energy Works', *Future World Comics*, 1 (New York: Summer 1946)
Komunyakaa, Yusef, '1984', *Callaloo*, 20 (Winter, 1984), 114–118
McCammon, Robert R., *Swan Song* (New York: Pocket Books, 1987)
McGrann, Molly, 'Three Mile Island', *Columbia: A Journal of Literature and Art*, 30 (Fall 1998)
Oliphant, Pat, 'Chernobyl!!', April 29, 1986, in Pat Oliphant and Susan Conway, *The New World Order in Drawing and Sculpture 1983–1993* (Kansas: Andrews and McMeel, 1994)
—— 'Rockets', in *The New World Order in Drawing and Sculpture* (Missouri: Universal Press Syndicate Company, 1994)
Picture News, 1.1. (New York: January 1946)
Rhinehart, Luke, *Long Voyage Back* (London: Grafton, 1983)
Science Comics, 1, (Springfield: Humor Publications, January 1946)
Sisson, Jonathan, 'The Crows of St Thomas', *Poetry*, 139.3 (December 1981), 139–141
'Starbound Rocket', *True Comics*, 54 (New York: November 1946)
Strieber, Whitley and James Kunetka, *Warday* (New York: Henry Holt & Co, 1984)
Swanwick, Michael, *In The Drift* (New York: Ace Science Fiction Books, 1985)
'Untitled Karen Silkwood Comic by R. Diggs', *Corporate Crime Comics*, I (July 1977)
Wells, H. G., *The World Set Free* (London: Collins, 1956)
Wentzell, Timothy, *Faded Giant* (Bloomington: AuthorHouse, 2009)

Music, Radio, Theatre, and Podcast

'The 1970s', *Letters from America*, BBC, c. 1970s, available online at www.bbc.co.uk/programmes/b03zj367 (accessed April 14, 2016)
The Buchanan Brothers, *Atomic Power* (Atomic Platters: Cold War Music from the Golden Age of Homeland Security, 2010)
Hoffmann, Karl, Guido Barbieri, Andrea Molino and Oscar Pizzo, *The Three Mile Island Opera*, ZKM | Karlsruhe, March 31, 2012
Kirby, Fred, *When That Hell Bomb Falls* (Atomic Platters: Cold War Music from the Golden Age of Homeland Security, 2010)
Lehrer, Tom, *We'll All Go Together When We Go* (Lehrer Records, 1959)
The Louvin Brothers *Great Atomic Power* (Atomic Platters: Cold War Music from the Golden Age of Homeland Security, 2010)
The Sons of the Pioneers, *Old Man Atom* (Atomic Platters: Cold War Music from the Golden Age of Homeland Security, 2010)

Images Used

'Atoms Unleashed: The Story of the Atomic Bomb', *True Comics*, 47 (New York: March 1946)

'Crowd at rally. Anti-nuke rally in Harrisburg (Pennsylvania) at the Capitol', President's Commission on the Accident at Three Mile Island, March 29–April 30, 1979, in *National Archives Catalog* (n.d.), available online at https://research.archives.gov/id/540017 (accessed September 22, 2016)

'Demonstrators' signs. Anti-nuke rally in Harrisburg, (Pennsylvania) at the Capitol', President's Commission on the Accident at Three Mile Island, March 29–April 30, 1979, 3/29/1979–4/30/1979, in *National Archives Catalog* (n.d.), available online at https://research.archives.gov/id/540020 (accessed September 22, 2016)

Kelton, Dan, *The Dispatch*, 22 January 1982, in James. A. Ollinger, 'The Dispatch' (n.d.), available online at www.jollinger.com/dispatch/ (accessed July 17, 2016)

Mantlo, Bill, Gene Colan and Dave Simons, 'A Christmas for Carol', *Howard the Duck*, 1.3 (New York: February 1980)

'Mushroom Cloud', Department of Defense. Department of the Air Force, in *National Archives Catalog* (n.d.), available online at https://research.archives.gov/id/542192 (accessed September 22, 2016) ARC Identifier 542192

'President Jimmy Carter leaving (Three Mile Island) for Middletown, Pennsylvania', President's Commission on the Accident at Three Mile Island, March 29–April 30, 1979, in *National Archives Catalog* (n.d.), available online at https://catalog.archives.gov/id/540021 (accessed September 22, 2016) ARC Identifier: 540021

United States Atomic Energy Commission (1955), in Jean-Marc Wolff, 'Eurochemic (1956–1990)', Nuclear Energy Agency (1996), available online at www.eurochemic.be/nl/documents/68-eurochemic-EN.pdf (accessed September 22, 2016), p.52

'Untitled: HD.3C.029', in *Energy.Gov* (2013), available online at https://www.flickr.com/photos/departmentofenergy/10692186803/ (accessed September 22, 2016)

X-Men Origins: Wolverine, dir. by Gavin Hood (20th Century Fox, 2009)

Souvenirs

The Original Canned Radiation, Brenster Enterprises of Etters (Pennsylvania, c. 1980)

Applications

Nuclear Plants, developed by Claus Zimmerman, October 22, 2012, version 1.3

Nuclear Power Plants, developed by Michael Hoereth, October 10, 2013, version 2.7

Nuclear Sites Free, developed by Janak Shah, March 29, 2012, version 1.3.1

Nuclear Tycoon, developed by Justin Lehmann, May 17, 2016, vision 1.0.1

Radiation Map Tracker, developed by Black Cat Systems, July 1, 2014, version 1.2

Index

Page numbers in *italics* refer to notes.